AF360810

3D Analysis of Human Movement, Sport, and Health Promotion

3D Analysis of Human Movement, Sport, and Health Promotion

Editor

Luca Petrigna

Basel • Beijing • Wuhan • Barcelona • Belgrade • Novi Sad • Cluj • Manchester

Editor
Luca Petrigna
Scienze Biomediche e
Biotecnologiche
University of Catania
Catania
Italy

Editorial Office
MDPI
St. Alban-Anlage 66
4052 Basel, Switzerland

This is a reprint of articles from the Special Issue published online in the open access journal *Journal of Functional Morphology and Kinesiology* (ISSN 2411-5142) (available at: https://www.mdpi.com/journal/jfmk/special_issues/3D_Human_Movement).

For citation purposes, cite each article independently as indicated on the article page online and as indicated below:

Lastname, A.A.; Lastname, B.B. Article Title. *Journal Name* **Year**, *Volume Number*, Page Range.

ISBN 978-3-0365-9741-6 (Hbk)
ISBN 978-3-0365-9740-9 (PDF)
doi.org/10.3390/books978-3-0365-9740-9

Contents

Journal of
Functional Morphology and Kinesiology

Editorial

3D Analysis of Human Movement, Sport, and Health Promotion

Luca Petrigna * and Giuseppe Musumeci

Department of Biomedical and Biotechnological Sciences, Section of Anatomy, Histology and Movement Science, School of Medicine, University of Catania, Via S. Sofia 97, 95123 Catania, Italy; giuseppe.musumeci@unict.it
* Correspondence: luca.petrigna@unict.it

Citation: Petrigna, L.; Musumeci, G. 3D Analysis of Human Movement, Sport, and Health Promotion. *J. Funct. Morphol. Kinesiol.* **2023**, *8*, 157. https://doi.org/10.3390/jfmk8040157

Received: 23 October 2023
Accepted: 3 November 2023
Published: 7 November 2023

This Special Issue, "3D Analysis of Human Movement, Sport, and Health Promotion", aimed to collect studies that assessed motor functions and alterations. The idea was to focus attention on objective and quantitative evaluation methods in both sport and, primarily, health promotion. According to the World Health Organization, health is a "complete physical, mental and social well-being, and not merely the absence of disease or infirmity" [1]. It is consequently important to think about well-designed health interventions in this way, to reduce the risks of correlated health status problems and reduce costs in healthcare systems [2,3]. On the one hand, the education of the population is fundamental for better knowledge on healthy habits; on the other hand, it is important to perform regular screening to evaluate health status [4,5].

It is possible to evaluate health status with laboratory tests (which usually present higher reliability and validity) and field tests (which are cheaper and easier to administer) [6]. Usually, field tests are more ecologically valid, making them a good solution for population-based studies [7]. Both solutions have to be investigated to achieve high-quality evaluations, as well as tests that can be performed involving more people in less time. One aspect that laboratory and field evaluations should have in common is the scientific quality of the protocol. Recently, the concept of the standard operating procedure in physical fitness evaluation has been proposed [8]. The idea of this concept is a detailed step-by-step description of the protocol to increase the quality of the research, allow the repetition of the protocol, and compare the data [8]. It is a concept adopted in other disciplines such as in the fields of medicine and engineering, where errors are not allowed [8]; consequently, it should also be suitable for the field of health evaluation.

The importance of evaluation with three-dimensional (3D) analysis is that it allows a vision of the human body on the three planes to be obtained, increasing the possibility of better understanding all aspects that could bring functional alterations [9]. The idea behind 3D testing is that kinematic analysis, especially if performed in 3D, helps in the evaluation of the quality of movements [10]. This is extremely helpful in the rehabilitation setting, as the study of the human kinematics for movement quantification is fundamental for evaluating improvements in patients, such as after a stroke [11,12], in scoliosis evaluation [13], or for digital postural analysis [14].

The articles included in this Special Issue address the 3D evaluation of human movement in different populations, the practical application of postural analysis, and possible future directions of research in this field. Different methods could be adopted in 3D evaluation, such as multicamera systems, machine learning approaches, or specific software for computers and smartphones. One of the studies included in this Special Issue adopted four infrared 3D cameras using the SMART Integrated System [15]. Russo and colleagues, in their article [15], evaluated the impact of Nordic walking pole length on healthy participants. The authors of [15] adopted a technique that is considered the gold standard for dynamic evaluation, but the price of the instrument limits its use. An alternative method detected in the included studies in the present Special Issue for human 3D evaluation is photogrammetry. It is a radiation-free, easy, inexpensive, and rapid tool that can help in postural screening and repetitive controls. It was adopted in the study by Belli and colleagues, who suggest that this instrument works especially well in bending evaluation [16].

The above systems are not always accessible for all clinicians due to their cost and the skills required to use them, suggesting the use of more economical techniques such as the calculation of body angles from photographs or the use of goniometry may be more attainable [17]. Following this principle, the study in [18] adopted software that evaluated human posture after the positioning of markers on specific anatomical landmarks. The authors in [18] analyzed body posture accurately and the software adopted was Dartfish ProSuite 6 (Dartfigh, Fribourg, Switzerland). Another study [19] adopted a mobile app named Apecs (Apecs-AI Posture Evaluation and Correction System®. New Body Technologies SAS, Grenoble, France). This application demonstrated good reproducibility with trunk inclination, although axillae alignment was unreliable in all the planes. This evaluation is cheap, easy to adopt (the application indicates the anatomical points on which to place the markers), and feasible; consequently, it could be an alternative to more expensive devices.

Software recording with 3D systems sometimes requires the use of technology to analyze data. In the life sciences, there is an ever-increasing interest in machine learning [20]. This technology is based on deep learning and has multiple field applications; one that could be of interest for movement analysis is the identification of objects in images [21]. Machine learning tools are useful for analyzing 3D movement kinematics and they can help distinguish healthy from pathological behaviors [10]. Deep learning could be adapted to analyze behaviors with standardized methods proposed in the literature [22]. In one of the studies included in this Special Issue, a machine learning approach was adopted in the evaluation of climbing holding time [23]. The concept of the study could also be adopted in other studies using 3D data.

All the above aspects should be related with their practical applications. Using 3D analysis is not only useful for researchers but also for kinesiologists. Three-dimensional evaluation and analysis could be useful in a studio, in a sports setting, or in a rehabilitation setting, but could also be performed remotely. Indeed, new technologies, if integrated with proper applications and supported by a good internet connection, could allow for the monitoring of treatment, as was demonstrated in another study included in this Special Issue [18]. As a review suggests [24], everything proposed above could be moved to the metaverse, a place where these new technologies and evaluation techniques can be adopted to reach people everywhere and at every time.

A practical application proposed in another study is the evaluation of hand-standing from a posturographic point of view [25]. It is not a natural human position, but different sports and disciplines require this specific and not-so-often-studied posture. Consequently, the study provides feedback for future research on this topic. Future research could also consider the evaluation of the human posture by integrating the above-presented tests with other techniques such as a termocamera [14] or in dual-task conditions [26]. In this way, it is possible to increase the quality and accuracy of the evaluation.

This Special Issue seeks to provide an overview of instruments, research ideas, and also future applications of 3D analysis technology. In the near future, it will be interesting to make some of the above-presented instruments the gold standard or valid research tools that could complement much more expensive and complex instruments. Furthermore, researchers should start to think about adopting new markerless technologies or techniques to make research as objective, precise, and cheap as possible.

In conclusion, this editorial aimed to present an overview of the Special Issue "3D Analysis of Human Movement, Sport, and Health Promotion" and provide feedback for future studies on the same or similar topics.

Conflicts of Interest: The authors declare no conflict of interest.

References

1. World Health Organization. *Constitution of the World Health Organization*; WHO: Geneva, Switzerland, 1995.
2. Fries, J.F.; Harrington, H.; Edwards, R.; Kent, L.A.; Richardson, N. Randomized controlled trial of cost reductions from a health education program: The California Public Employees' Retirement System (PERS) study. *Am. J. Health Promot.* **1994**, *8*, 216–223. [CrossRef]
3. Galloway, R.D. Health promotion: Causes, beliefs and measurements. *Clin. Med. Res.* **2003**, *1*, 249–258. [CrossRef] [PubMed]
4. American College of Sports Medicine. The recommended quantity and quality of exercise for developing and maintaining cardiorespiratory and muscular fitness, and flexibility in healthy adults. *Med. Sci. Sports Exerc.* **1998**, *30*, 975–991.
5. Ruiz, J.R.; Castro-Piñero, J.; Artero, E.G.; Ortega, F.B.; Sjöström, M.; Suni, J.; Castillo, M.J. Predictive validity of health-related fitness in youth: A systematic review. *Br. J. Sports Med.* **2009**, *43*, 909–923. [CrossRef] [PubMed]
6. Heyward, V.H. Advanced fitness assessment and exercise prescription. *Med. Sci. Sports Exerc.* **1992**, *24*, 278. [CrossRef]
7. Artero, E.; Espana-Romero, V.; Castro-Pinero, J.; Ortega, F.; Suni, J.; Castillo-Garzon, M.; Ruiz, J. Reliability of field-based fitness tests in youth. *Int. J. Sports Med.* **2011**, *32*, 159–169. [CrossRef]
8. Petrigna, L.; Pajaujiene, S.; Delextrat, A.; Gómez-López, M.; Paoli, A.; Palma, A.; Bianco, A. The importance of standard operating procedures in physical fitness assessment: A brief review. *Sport Sci. Health* **2022**, *18*, 21–26. [CrossRef]
9. Roggio, F.; Ravalli, S.; Maugeri, G.; Bianco, A.; Palma, A.; Di Rosa, M.; Musumeci, G. Technological advancements in the analysis of human motion and posture management through digital devices. *World J. Orthop.* **2021**, *12*, 467–484. [CrossRef]
10. Arac, A. Machine learning for 3D kinematic analysis of movements in neurorehabilitation. *Curr. Neurol. Neurosci. Rep.* **2020**, *20*, 1–6. [CrossRef]
11. Alt Murphy, M.; Willén, C.; Sunnerhagen, K.S. Responsiveness of upper extremity kinematic measures and clinical improvement during the first three months after stroke. *Neurorehabil. Neural Repair* **2013**, *27*, 844–853. [CrossRef]
12. Kwakkel, G.; Lannin, N.A.; Borschmann, K.; English, C.; Ali, M.; Churilov, L.; Saposnik, G.; Winstein, C.; van Wegen, E.E.; Wolf, S.L.; et al. Standardized measurement of sensorimotor recovery in stroke trials: Consensus-based core recommendations from the Stroke Recovery and Rehabilitation Roundtable. *Int. J. Stroke* **2017**, *12*, 451–461. [CrossRef] [PubMed]
13. Roggio, F.; Petrigna, L.; Filetti, V.; Vitale, E.; Rapisarda, V.; Musumeci, G. Infrared thermography for the evaluation of adolescent and juvenile idiopathic scoliosis: A systematic review. *J. Therm. Biol.* **2023**, *113*, 103524. [CrossRef]
14. Roggio, F.; Petrigna, L.; Trovato, B.; Zanghì, M.; Sortino, M.; Vitale, E.; Rapisarda, L.; Testa, G.; Pavone, V.; Pavone, P.; et al. Thermography and rasterstereography as a combined infrared method to assess the posture of healthy individuals. *Sci. Rep.* **2023**, *13*, 4263. [CrossRef] [PubMed]
15. Russo, L.; Belli, G.; Di Blasio, A.; Lupu, E.; Larion, A.; Fischetti, F.; Montagnani, E.; Di Biase Arrivabene, P.; De Angelis, M. The Impact of Nordic Walking Pole Length on Gait Kinematic Parameters. *J. Funct. Morphol. Kinesiol.* **2023**, *8*, 50. [CrossRef] [PubMed]
16. Belli, G.; Toselli, S.; Mauro, M.; Maietta Latessa, P.; Russo, L. Relation between Photogrammetry and Spinal Mouse for Sagittal Imbalance Assessment in Adolescents with Thoracic Kyphosis. *J. Funct. Morphol. Kinesiol.* **2023**, *8*, 68. [CrossRef]
17. Fortin, C.; Ehrmann Feldman, D.; Cheriet, F.; Labelle, H. Clinical methods for quantifying body segment posture: A literature review. *Disabil. Rehabil.* **2011**, *33*, 367–383. [CrossRef]
18. Ludwig, O.; Dindorf, C.; Schuh, T.; Haab, T.; Marchetti, J.; Fröhlich, M. Effects of Feedback-Supported Online Training during the Coronavirus Lockdown on Posture in Children and Adolescents. *J. Funct. Morphol. Kinesiol.* **2022**, *7*, 88. [CrossRef]
19. Trovato, B.; Roggio, F.; Sortino, M.; Zanghì, M.; Petrigna, L.; Giuffrida, R.; Musumeci, G. Postural Evaluation in Young Healthy Adults through a Digital and Reproducible Method. *J. Funct. Morphol. Kinesiol.* **2022**, *7*, 98. [CrossRef]
20. Camacho, D.M.; Collins, K.M.; Powers, R.K.; Costello, J.C.; Collins, J.J. Next-Generation Machine Learning for Biological Networks. *Cell* **2018**, *173*, 1581–1592. [CrossRef]
21. LeCun, Y.; Bengio, Y.; Hinton, G. Deep learning. *Nature* **2015**, *521*, 436–444. [CrossRef]
22. Arac, A.; Zhao, P.; Dobkin, B.H.; Carmichael, S.T.; Golshani, P. DeepBehavior: A Deep Learning Toolbox for Automated Analysis of Animal and Human Behavior Imaging Data. *Front Syst. Neurosci.* **2019**, *13*, 20. [CrossRef] [PubMed]
23. Dindorf, C.; Bartaguiz, E.; Dully, J.; Sprenger, M.; Merk, A.; Becker, S.; Fröhlich, M.; Ludwig, O. Evaluation of Influencing Factors on the Maximum Climbing Specific Holding Time: An Inferential Statistics and Machine Learning Approach. *J. Funct. Morphol. Kinesiol.* **2022**, *7*, 95. [CrossRef] [PubMed]
24. Petrigna, L.; Musumeci, G. The metaverse: A new challenge for the healthcare system: A scoping review. *J. Funct. Morphol. Kinesiol.* **2022**, *7*, 63. [CrossRef] [PubMed]
25. Thomas, E.; Rossi, C.; Petrigna, L.; Messina, G.; Bellafiore, M.; Şahin, F.N.; Proia, P.; Palma, A.; Bianco, A. Evaluation of Posturographic and Neuromuscular Parameters during Upright Stance and Hand Standing: A Pilot Study. *J. Funct. Morphol. Kinesiol.* **2023**, *8*, 40. [CrossRef]
26. Petrigna, L.; Gentile, A.; Mani, D.; Pajaujiene, S.; Zanotto, T.; Thomas, E.; Paoli, A.; Palma, A.; Bianco, A. Dual-task conditions on static postural control in older adults: A systematic review and meta-analysis. *J. Aging Phys. Act.* **2020**, *29*, 162–177. [CrossRef]

Journal of
Functional Morphology and Kinesiology

Article

Postural Evaluation in Young Healthy Adults through a Digital and Reproducible Method

Bruno Trovato [1], Federico Roggio [1,2], Martina Sortino [1], Marta Zanghì [1], Luca Petrigna [1,*], Rosario Giuffrida [3] and Giuseppe Musumeci [1,4,5]

1   Department of Biomedical and Biotechnological Sciences, Section of Anatomy, Histology and Movement Science, School of Medicine, University of Catania, Via S. Sofia No. 97, 95123 Catania, Italy
2   Sport and Exercise Sciences Research Unit, Department of Psychology, Educational Science and Human Movement, University of Palermo, Via Giovanni Pascoli 6, 90144 Palermo, Italy
3   Department of Biomedical and Biotechnological Sciences, Section of Physiology, School of Medicine, University of Catania, 95125 Catania, Italy
4   Research Center on Motor Activities (CRAM), University of Catania, Via S. Sofia No. 97, 95123 Catania, Italy
5   Department of Biology, Sbarro Institute for Cancer Research and Molecular Medicine, College of Science and Technology, Temple University, Philadelphia, PA 19122, USA
*   Correspondence: luca.petrigna@unict.it

Abstract: Different tools for the assessment of posture exist, from the simplest and cheap plumb line to complex, expensive, 3D-marker-based systems. The aim of this study is to present digital postural normative data of young adults collected through a mobile app to expand the possibilities of digital postural evaluation. A sample of 100 healthy volunteers, 50 males and 50 females, was analyzed with the mobile app Apecs-AI Posture Evaluation and Correction System® (Apecs). The Student's *t*-test evaluated differences between gender to highlight if the digital posture evaluation may differ between groups. A significant difference was present in the anterior coronal plane for axillary alignment ($p = 0.04$), trunk inclination ($p = 0.03$), and knee alignment ($p = 0.01$). Head inclination ($p = 0.04$), tibia shift ($p = 0.01$), and foot angle ($p < 0.001$) presented significant differences in the sagittal plane, while there were no significant differences in the posterior coronal plane. The intraclass correlation coefficient (ICC) was considered to evaluate reproducibility. Thirteen parameters out of twenty-two provided an ICC > 0.90, three provided an ICC > 0.60, and six variables did not meet the cut-off criteria. The results highlight that digital posture analysis of healthy individuals may present slight differences related to gender. Additionally, the mobile app showed good reproducibility according to ICC. Digital postural assessment with Apecs could represent a quick method for preventing screening in the general population. Therefore, clinicians should consider this app's worth as an auxiliary posture evaluation tool.

Keywords: posture; reproducibility; mobile app; movement; kinesiology

Citation: Trovato, B.; Roggio, F.; Sortino, M.; Zanghì, M.; Petrigna, L.; Giuffrida, R.; Musumeci, G. Postural Evaluation in Young Healthy Adults through a Digital and Reproducible Method. *J. Funct. Morphol. Kinesiol.* **2022**, *7*, 98. https://doi.org/10.3390/jfmk7040098

Academic Editor: Vítor P. Lopes

Received: 12 October 2022
Accepted: 27 October 2022
Published: 28 October 2022

Publisher's Note: MDPI stays neutral with regard to jurisdictional claims in published maps and institutional affiliations.

1. Introduction

Posture is defined as the position acquired by the human body in various situations, opposed to the force of gravity and adapting to different environments [1]. Moreover, posture is essential for maintaining postural balance both in static and dynamic conditions. Remaining in a non-ergonomic position for an extended period can predispose people to manifest musculoskeletal pain [2]; thus, assuming good postures is considered necessary for general health at both a musculoskeletal and psychological level [3]. Nowadays, the evaluation of human posture is performed consistently in healthcare clinics and fitness centers, considering that postural misalignments can cause individuals to manifest headaches, lower back pain, neck pain, neurological pathologies, and a reduction in overall psychological well-being [4]. Currently, the literature does not provide any gold standard procedure for postural assessment. The types of exams employed can vary from a visual evaluation with goniometers and plumb lines to motion capture systems, such as Vicon, for dynamic evaluation and 3D camera infrared systems, such as rasterstereography, for static evaluation [5]. Regarding the feasibility of using a markerless system to assess human posture, rasterstereography is a system that generates a 3D model of the spine by calculating specific deformities and analyzing the convexity and concavity of the spine [6]. It is commonly used to investigate the presence of scoliosis and is considered reliable for the assessment of parameters such as pelvic obliquity, thoracic kyphosis, and lumbar lordosis angles [7,8]. However, this system has a high cost, and it is difficult to implement in postural screening for the general population. Other valid tools such as inertial measurement units (e.g., accelerometers, magneto inertial units) are also employed in the field of postural evaluation for the assessment of the thoracic kyphosis and the lumbar lordosis angles [9] and also for gait and balance assessment [10].

All the available methods for evaluating posture present some biases or disadvantages. The visual evaluation with a plumb line is cheap, but it requires specialized personnel, is prone to bias, and lacks scientific validation [11]. The use of goniometers is feasible for the measurement of the range of motion and angles of different joints with good reliability [12]; it has a low cost and is easy to perform, although it presents some methodological issues when assessing postural deviations [13], and it is only considered useful for one postural variable examination at the time [5]. Marker-based advanced technologies that can provide highly accurate data on joint angles and translations are potentially available for clinicians; however, these evaluation systems are too expensive for the average clinic, and often they are employed for research purposes only [14].

In this heterogeneous scenario regarding the available postural evaluation tools, the advancement in image-based technologies will come in handy for clinicians and researchers who want to find a postural assessment system with good reproducibility and an affordable cost. Tablet and phone apps for postural evaluation can fill this gap, with different postural apps demonstrating promising results in the evaluation of the frontal plane [15], standing posture [13], angulation variables [14], and head shift in sagittal and frontal planes [4]; however, the literature is insufficient to confirm the quality of these methods. Considering that the complete visual evaluation of body posture with goniometers and a plumb line can be long and not free from biases, and taking into account the high costs of 3D systems, the use of a mobile app could represent a quick, safe, and accurate method for researchers and clinicians to quantitatively evaluate general posture. Moreover, laboratory tests are often more expensive than field-based ones [16], and adopting a mobile, affordable tool for postural assessment could benefit the primary prevention of musculoskeletal disorders of the spine. The aim of this study is to present normative data about digital posture evaluation collected through a mobile app Apecs and, moreover, to evaluate the reproducibility.

2. Materials and Methods

2.1. Participants

We recruited and evaluated a sample of 100 healthy volunteers, 50 males and 50 females, with a mean age of 23.4 (standard deviation (SD) $\pm$ 6.2) years. Prior to testing, all participants were informed about the study procedure, risks, and benefits and provided written, informed consent to participate in the study and for us to use their data. The study followed the Helsinki Declaration principles and was approved by the University of Catania (protocol no.: CRAM-017-2020, 16 March 2020).

Exclusion criteria comprised: past or current major musculoskeletal injuries, spine pathology, and neurological pathologies. All the participants selected after the oral interview underwent a static postural evaluation provided by expert clinician (experience of 7 years) to confirm their eligibility for the study.

2.2. Study Design

The evaluation took 30 min per participant at the University Laboratories and consisted of evaluating their health status and history of previous conditions that could meet the exclusion criteria. After the screening process, participants were always asked to attend the laboratory at the same time (between 10:00 a.m. and 12:00 p.m.). The postural evaluation was performed by three different operators with similar experience in postural analysis (3 to 4 years of experience). The mobile app Apecs-AI Posture Evaluation and Correction System® (New Body Technologies SAS, Grenoble, France) (Apecs app) was used to acquire the images of the participants in standing position. The participants were asked to dress in minimal clothing, shorts for men and shorts and bras for women, to minimize biases relating to wrong landmark positioning during the postural analysis; for the same reason, markers were placed by expert clinicians on the body of the participants in correspondence to the app's predetermined landmarks. Four pictures were captured, one for the anterior coronal plane, one for the posterior coronal plane, and two for the sagittal plane (left and right). Participants were instructed to place their feet at the same width as the shoulders (Figure 1).

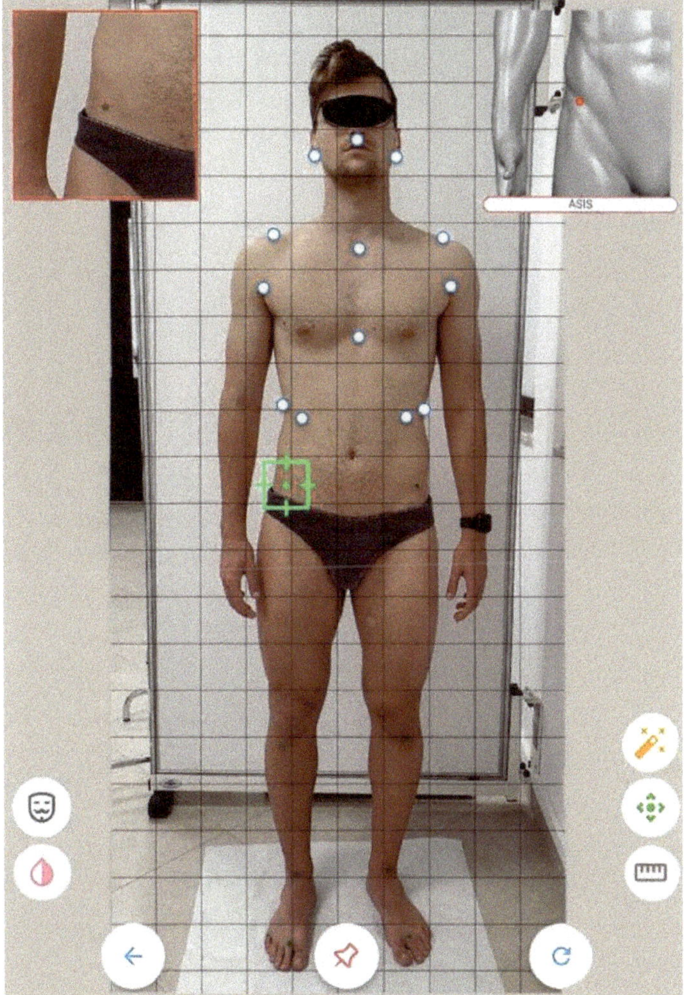

Figure 1. Landmarks positioning.

To avoid any wrong camera leveling during the image acquisition, the app's interface shows a target that becomes green when the camera is leveled. After the picture is acquired, the app immediately steers the user to crop the image at the individual's head and feet to minimize inconsistency in the proportion of different images. The Apecs app uses standardized digital landmarks and anatomical angles from one to four pictures, depending on the number of variables of interest to the investigation. The app calculates 24 postural variables from the predetermined anatomical markers in the three planes of the space examined. Figure 2 shows the points evaluated in the anterior coronal plane (a), the sagittal plane (b), and the posterior coronal plane (c).

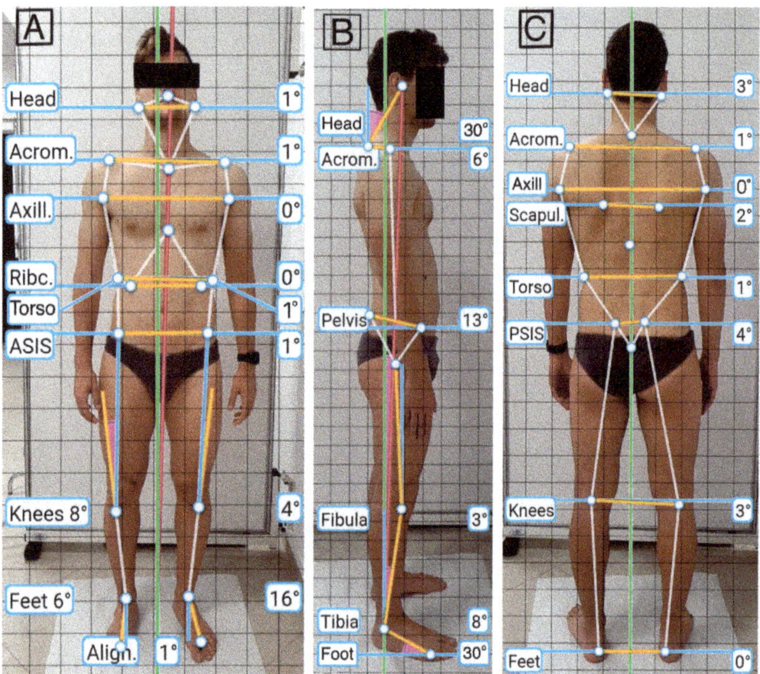

Figure 2. Evaluation of the anterior coronal plane (**A**); of the sagittal plane (**B**); of the posterior coronal plane (**C**).

After the cropping phase, the app drives the user to position the digital markers, fostering this process with examples of the proper positioning with images. Table 1 shows all the anatomical landmarks taken into consideration by the app for calculating the postural variables.

Table 1. Anatomic landmarks and postural variables studied with Apecs.

Plane of the Space	Anatomical Landmarks	Postural Variables
Anterior coronal	Acromion Anterior axillary folds Anterior superior iliac spine Jugular notch Lobulus auriculae Lowest point of costal margin Midpoint between malleoli Most intended point of the trunk Philtrum Second metatarsophalangeal joint Tibial tuberosity Xiphoid process	Body alignment Head alignment Acromion alignment Axillae alignment Trunk inclination Ribcage tilt Antero superior iliac spine inclination Knee angle
Posterior coronal	Lobulus auriculae C-7 vertebrae Acromion Anterior axillary folds Inferior angle of the scapula T-6 vertebrae Most intended point of the trunk Posterior superior iliac spine Superior end of intergluteal cleft Popliteal fossa Calcaneal tuberosity	Body alignment Head alignment Shoulder alignment Axillae alignment Scapulae alignment Trunk inclination Postero superior iliac spines Knee angle Foot angle
Sagittal	Tragus C-7 vertebrae Acromion Posterior superior iliac spine Greater trochanter Lateral joint line Lateral malleolus Head of the fifth metatarsal bone	Body alignment Head alignment Acromion alignment Pelvic tilt Tibia shift Fibula alignment Foot angle

2.3. Statistical Analysis

Data analysis comprised descriptive statistics to present the mean and standard deviation of the whole sample divided by gender. Inferential statistics comprised the Shapiro–Wilk test to assess the data distribution; the Student's *t*-test was used to compare means between the male and female groups; statistical significance was set at $p \leq 0.05$. Cohen's effect size (d) was applied to identify meaningful differences between the groups. Based on Cohen's criteria, d = 0.80 (absolute value) was considered a large effect size, and d = 0.50 (absolute value) was considered a medium effect size. Post hoc power calculations were performed with G*Power v.3.1. Three qualified examiners were selected to perform the positioning of the markers and the postural analysis in the two different parts of the day to assess the reproducibility of the app. The two-way mixed effect for absolute agreement was the model for calculating the intraclass correlation coefficient (ICC) for inter-rater agreement. The cut-off values for reproducibility based on a 95% confidence interval of the ICC estimate were <0.5 (poor), between 0.5 and 0.75 (moderate), between 0.75 and 0.9 (good), and >0.9 (excellent) [17]. All the statistical analyses were performed with R Project for Statistical Computing (Vienna, Austria).

3. Results

Anthropometric measurements were taken for each subject and grouped by gender, with a mean male height of 175 (SD ± 5.6) cm, a mean female height of 164.6 (SD ± 6.5) cm, and a mean male weight of 75.5 (SD ± 8.8) kg and a mean female weight of 58.13 (SD ± 7.41) kg.

The post hoc power calculation analysis with G*Power 3.1 returned a statistical power of 0.696 for our sample. The analysis of the digital anatomical landmarks collected with the Apecs app and the ICC values are presented in Table 2. The Student's *t*-test statistically indicated differences in the postural evaluation performed by the mobile app Apecs between males and females for specific variables. The postural variables with significant differences between male and female groups in the anterior coronal plane were axillary alignment ($p = 0.04$), trunk inclination ($p = 0.03$), and knee alignment ($p = 0.01$). The female group presented more body inclination to the right than men, more trunk inclination, and a wider knee angle in the anterior coronal plane. The male group showed the worst results for the axillary alignment, which resulted in greater deviation from the ideal alignment than that of the female group. In the sagittal plane, statistically significant differences were found for head inclination ($p = 0.04$), tibia shift ($p = 0.01$), and foot angle ($p < 0.001$). The head of the female group was more significantly shifted from the ideal alignment compared to that of the male group and also showed a more accentuated anterior tibial shift. Instead, the male group presented a wider foot angle than the female group. No statistically significant differences were found between groups for the evaluation of the posterior coronal plane. According to Cohen's d, there was a small effect size only for ribcage tilt ($d = -0.35$) in the anterior coronal plane, for head alignment in the sagittal plane ($d = -0.38$), and a large effect size for knee angle in the anterior coronal plane ($d = -0.89$), tibia shift in the sagittal plane ($d = -0.95$), and foot angle in the sagittal plane ($d = 1.6$).

Figures 3–5 show the box plots for gender differences in the three space planes.

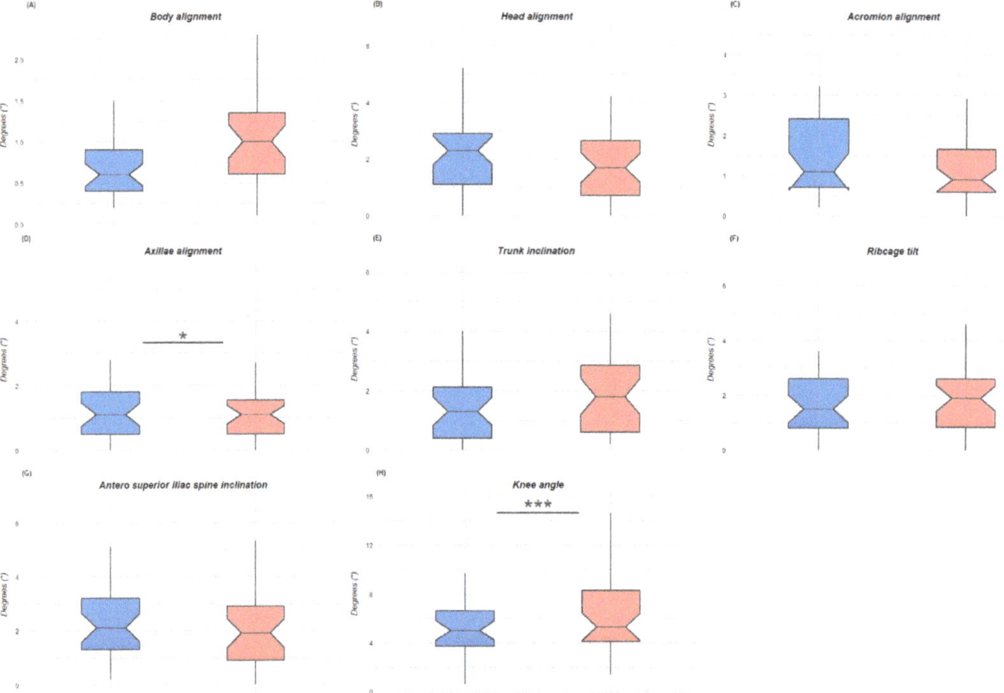

Figure 3. Box plots of the differences between male and female groups in the anterior coronal plane with indication of significance. Figure (**A**) is for body alignment; figure (**B**) is for head alignment, figure (**C**) is of acromion alignment, figure (**D**) is for axillae alignment, figure (**E**) is for trunk inclination, figure (**F**) is for ribcage tilt, figure (**G**) is for antero superior iliac spine inclination, figure (**H**) is for knee angle. *: $p < 0.05$; ***: $p < 0.001$.

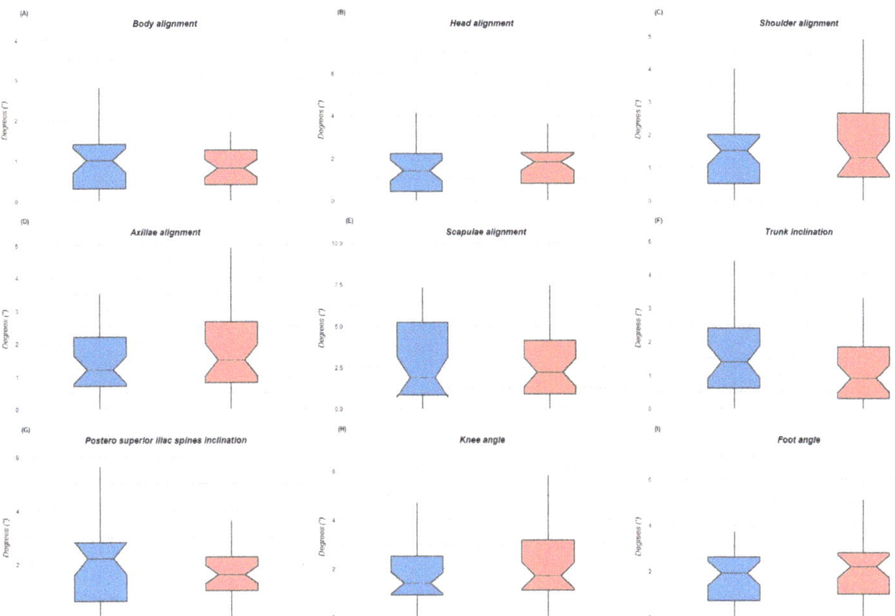

Figure 4. Box plots of the differences between male and female groups in the posterior coronal plane. Figure (**A**) is for body alignment; figure (**B**) is for head alignment, figure (**C**) is of shoulder alignment, figure (**D**) is for axillae alignment, figure (**E**) is for scapulae alignment, figure (**F**) is for trunk inclination, figure (**G**) is for postero superior iliac spine inclination, figure (**H**) is for knee angle, figure (**I**) is for foot angle.

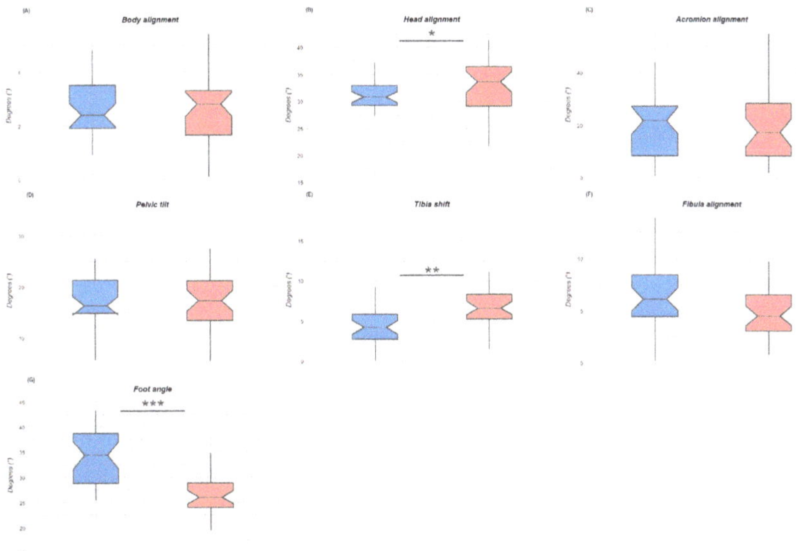

Figure 5. Box plots of the postural differences between male and female groups in the sagittal plane with indication of significance. Figure (**A**) is for body alignment; figure (**B**) is for head alignment, figure (**C**) is of acromion alignment, figure (**D**) is pelvic tilt, figure (**E**) is for tibia shift, figure (**F**) is for fibula alignment, figure (**G**) is for foot angle. *: $p < 0.05$; **: $p < 0.01$; ***: $p < 0.001$.

Table 2. Description of group means and ICC of the postural variables analyzed.

	Postural Variables	Total	Males	Females	*t*-Test	ICC	Cohen d
		Mean ± SD	Mean ± SD	Mean ± SD			
Anterior coronal	Body alignment	0.9° ± 0.5	0.7° ± 0.4	1° ± 0.5	0.430	0.95	−0.54
	Head alignment	2° ± 1.4	2.2° ± 1.8	1.8° ± 1.4	0.989	0.51	0.25
	Acromion alignment	1.3° ± 1	1.4° ± 0.9	1.2° ± 1.1	0.423	0.91	0.17
	Axillae alignment	1.3° ± 1	1.4° ± 1.4	1.2° ± 0.8	0.044 *	0.25	0.16
	Trunk inclination	1.6° ± 1.2	1.4° ± 1.3	1.8° ± 1.1	0.462	0.44	−0.29
	Ribcage tilt	1.9° ± 1.6	1.7° ± 1.2	2.2° ± 1.8	0.039 *	0.93	−0.35
	ASIS inclination	2.3° ± 1.6	2.5° ± 1.7	1.5° ± 0.3	0.321	0.94	0.24
	Knee angle	6.2° ± 3.3	4.8° ± 2.9	7.5° ± 3.1	0.001 ***	0.93	−0.89
Posterior coronal	Body alignment	1° ± 0.8	1° ± 0.8	0.9° ± 0.8	0.717	0.84	0.07
	Head alignment	2.7° ± 1.5	1.6° ± 1.4	1.8° ± 1.5	0.652	0.30	−0.14
	Shoulder alignment	1.5° ± 1.2	1.4° ± 1.1	1.7° ± 1.4	0.444	0.93	−0.19
	Axillae alignment	1.6° ± 1.2	1.4° ± 1	1.7° ± 1.2	0.348	0.43	−0.26
	Scapulae alignment	2.7° ± 2.2	2.8° ± 2.3	2.7° ± 2.2	0.879	0.92	0.05
	Trunk inclination	1.4° ± 1.2	1.6° ± 1.2	1.3° ± 1.2	0.925	0.26	0.23
	PSIS inclination	1.9° ± 1.4	2° ± 1.6	1.8° ± 1.3	0.247	0.66	0.15
	Knee angle	1.9° ± 1.4	1.8° ± 1.2	2.1° ± 1.6	0.172	0.94	−0.23
	Foot angle	2.1° ± 1.6	1.9° ± 1.4	2.3° ± 1.8	0.151	0.75	−0.24
Sagittal	Body alignment	2.6° ± 1.2	2.6° ± 1.1	2.5° ± 1.3	0.691	0.94	0.1
	Head alignment	31.4° ± 5.4	30.3° ± 4.3	32.4° ± 6.2	0.047 *	0.91	−0.38
	Acromion alignment	19.6° ± 12.3	19.9° ± 12.1	19.4° ± 12.8	0.866	0.24	0.04
	Pelvic tilt	16.9° ± 5.7	16.6° ± 5.3	17.1° ± 6.1	0.763	0.94	−0.07
	Tibia shift	5.7° ± 3.3	4.2° ± 2.2	7.1° ± 3.5	0.017 **	0.91	−0.95
	Fibula alignment	5.5° ± 3	6.3° ± 3.1	4.8° ± 2.7	0.491	0.94	0.49
	Foot angle	29.8° ± 6	33.7° ± 26.2	26.2° ± 4.2	0.001 ***	0.93	1.6

ASIS: anterior superior iliac spines; PSIS: postero superior iliac spines. * *p*-value < 0.05; ** *p*-value < 0.01; *** *p*-value < 0.001.

The ICC showed promising results for inter-rater reproducibility, with values > 0.90 for thirteen out of the twenty-two postural variables examined and >0.60 for the other three variables; only six variables did not meet the cut-off criteria required to be considered reliable. Table 2 shows the ICC for the postural variables evaluated.

4. Discussion

This study aimed to present normative data about the digital posture evaluation of healthy young adults performed by the mobile app Apecs and to evaluate its reproducibility. The first finding was that the app is sensible to postural variation, considering that it was capable of detecting postural differences between males and females. The second finding of the study was that this mobile app presents a good inter-rater reproducibility for all the postural variables examined except for head alignment, trunk inclination and axillae alignment in the anterior and posterior coronal plane, and acromion alignment in the sagittal plane.

The Apecs app has already been used for research purposes to evaluate postural behaviors related to specific ergonomic studies' work [12] and to evaluate body segment angles in subjects with adolescent idiopathic scoliosis [18]. However, the studies mentioned above had small samples; the first used the app only to compare their sample's posture at rest and during working activity, and the second only evaluated angles in the frontal and sagittal plane. Hence, to the best of our knowledge, this is the first study that employs the mobile app Apecs to evaluate global posture, providing normative data and assessing its reproducibility as a posture evaluation tool.

The sample in this study was composed of 100 participants equally distributed between males and females, and the Apecs mobile app was capable of detecting postural differences when present. It emerged from the postural analysis of the anterior coronal plane that females presented a wider knee angle; this could be due to the overall increased knee laxity and reduced stiffness in females compared to males [19]. In a previous study by Raine et al. [20], no sex differences were found for head inclination on the sagittal plane; conversely, we found that the head inclination was more accentuated in the female group compared to the male group. However, Raine et al.'s study dates from 1997, and they considered an older sample size. These observations may be the cause of the differences compared to our study. Iacob et al. [21] analyzed the posture of a sample of people with malocclusion through the PostureScreen® mobile app, comparing it with a healthy sample. We found a difference between our postural data gathered with Apecs and those reported by Iacob et al. for the same variable analyzed. These authors found, on the frontal plane, a head alignment in their sample of $3.86° \pm 2.45$, a shoulder alignment of 1 ± 0.97, and a hip deviation of 1.42 ± 1.28, while, for the same variable, we reported a head alignment of 2.7 ± 1.5, a shoulder alignment of 1.5 ± 1.2, and postero superior spine inclination of $1.9° \pm 1.4$. The differences in the postural evaluation between the two apps might be due to the differences in the samples considering that the control group of young healthy young adults investigated by Iacob et al. was composed of only 14 people, and they were almost exclusively females.

We found a statistically significant difference in the sagittal plane for foot angle, with the male group presenting higher values; this finding could be related to the generally bigger size of the foot anthropometrics of males [22]. In the anterior and posterior coronal planes, we did not find any statistically significant difference between gender for foot posture parameters, in line with previous studies [23,24].

The reproducibility analysis of Apecs showed excellent results for all the variables examined on the sagittal plane except for the acromion alignment. The marker placed on the acromion was not clearly visible during the positioning of the digital marker in this plane of space, making it difficult to evaluate with consistency among raters. The same issue occurred in the posterior coronal plane for the trunk inclination, where the app asks raters to identify "the most intended point of the trunk", which was not easy to replicate for the raters. Interestingly, the two less reliable measures in the anterior coronal plane were the axillae alignment and the trunk inclination, indicating that this mobile app should be carefully considered when a precise measure of these variables is needed. Accordingly, with what was stated by Szucs et al. [14] when evaluating the PostureScreen® mobile app, we suggest that the quality of the evaluation is higher when markers are placed on the subject and are clearly visible during the positioning of the digital markers; however, neither the Apecs manufacturer nor the PostureScreen mobile one specifies this in their instruction for postural analysis.

The current study presents some limitations. First, we considered a sample composed exclusively of young adults, so we could not assess whether the Apecs app is a feasible tool to employ in the postural evaluation of pediatric and elderly populations. Second, all the individuals in the sample were healthy, thus, these results should be carefully interpreted when examining individuals with pathologies that influence the musculoskeletal system. Third, we did not compare measures collected with Apecs with those collected by postural gold standard instruments to assess the validity of the app. Further studies should investigate the validity of Apecs as a reliable postural assessment tool, comparing it with rasterstereography or marker-based systems. However, these normative data may help those involved in the analysis of postural alterations as a comparative standard with a healthy sample. Finally, the digital landmark positioning accomplished with the app may be challenging for less experienced users and might change the evaluation results.

5. Conclusions

The mobile postural app Apecs demonstrated good reproducibility for most of the postural variables analyzed and could detect postural differences between males and females when present. The app was easy to use for all the raters, from the more experienced to the less experienced ones, indicating that Apecs could be a cheap and feasible good alternative to more expensive postural assessment devices for researchers and clinicians. However, trunk inclination and axillae alignment were unreliable in all the planes of space where they were evaluated, and head alignment was reliable only in the sagittal plane. Clinicians should be aware of this issue when using Apecs and carefully predetermine the landmark positioning and digital identification during the analysis to minimize the possibilities of errors for the postural variables not clearly described by the Apecs manufacturer. In conclusion, the Apecs app could be a potentially useful tool for clinicians and researchers to implement in the preventive care of postural disorders given its ease of use and cheap costs.

Author Contributions: Conceptualization, B.T. and F.R.; Methodology, B.T., F.R., and L.P.; Statistical Analysis, F.R., B.T., and R.G.; Investigation, G.M.; Resources, G.M.; Data Curation, M.S. and M.Z.; Writing—Original Draft Preparation, B.T., F.R., and L.P.; Writing—Review and Editing, R.G. and G.M.; Visualization, M.S. and M.Z. Supervision, G.M.; Funding Acquisition, G.M. All authors have read and agreed to the published version of the manuscript.

Funding: This work was supported by the University Research Project Grant PIACERI Found—NATURE-OA—2020–2022, Department of Biomedical and Biotechnological Sciences (BIOMETEC), University of Catania, Italy.

Institutional Review Board Statement: The study was conducted according to the guidelines of the Declaration of Helsinki and approved by the scientific committee of the University of Catania (protocol no.: CRAM-017-2020, 16 March 2020).

Informed Consent Statement: All subjects gave their informed consent for inclusion before they participated in the study.

Data Availability Statement: Data are available upon appropriate request.

Conflicts of Interest: The authors declare no conflict of interest.

References

1. Carini, F.; Mazzola, M.; Fici, C.; Palmeri, S.; Messina, M.; Damiani, P.; Tomasello, G. Posture and posturology, anatomical and physiological profiles: Overview and current state of art. *Acta Biomed.* **2017**, *88*, 11–16. [CrossRef]
2. Rodrigues, M.S.; Leite, R.D.V.; Lelis, C.M.; Chaves, T.C. Differences in ergonomic and workstation factors between computer office workers with and without reported musculoskeletal pain. *Work* **2017**, *57*, 563–572. [CrossRef]
3. Harvey, R.H.; Peper, E.; Mason, L.; Joy, M. Effect of Posture Feedback Training on Health. *Appl. Psychophysiol. Biofeedback* **2020**, *45*, 59–65. [CrossRef]
4. Hopkins, B.B.; Vehrs, P.R.; Fellingham, G.W.; George, J.D.; Hager, R.; Ridge, S.T. Validity and Reliability of Standing Posture Measurements Using a Mobile Application. *J. Manip. Physiol. Ther.* **2019**, *42*, 132–140. [CrossRef]
5. Fortin, C.; Feldman, D.E.; Cheriet, F.; Labelle, H. Clinical methods for quantifying body segment posture: A literature review. *Disabil. Rehabil.* **2011**, *33*, 367–383. [CrossRef]
6. Roggio, F.; Ravalli, S.; Maugeri, G.; Bianco, A.; Palma, A.; Di Rosa, M.; Musumeci, G. Technological advancements in the analysis of human motion and posture management through digital devices. *World J. Orthop.* **2021**, *12*, 467–484. [CrossRef]
7. Betsch, M.; Wild, M.; Jungbluth, P.; Hakimi, M.; Windolf, J.; Haex, B.; Horstmann, T.; Rapp, W. Reliability and validity of 4D rasterstereography under dynamic conditions. *Comput. Biol. Med.* **2011**, *41*, 308–312. [CrossRef]
8. Molinaro, L.; Russo, L.; Cubelli, F.; Taborri, J.; Rossi, S. Reliability analysis of an innovative technology for the assessment of spinal abnormalities. In Proceedings of the 2022 IEEE International Symposium on Medical Measurements and Applications (MeMeA), Messina, Italy, 22–24 June 2022; pp. 1–6.
9. Paloschi, D.; Bravi, M.; Schena, E.; Miccinilli, S.; Morrone, M.; Sterzi, S.; Saccomandi, P.; Massaroni, C. Validation and Assessment of a Posture Measurement System with Magneto-Inertial Measurement Units. *Sensors* **2021**, *21*, 6610. [CrossRef]
10. Leirós-Rodríguez, R.; Romo-Pérez, V.; García-Soidán, J.L.; Soto-Rodríguez, A. Identification of Body Balance Deterioration of Gait in Women Using Accelerometers. *Sustainability* **2020**, *12*, 1222. [CrossRef]
11. Tyson, S.F.; DeSouza, L.H. A clinical model for the assessment of posture and balance in people with stroke. *Disabil. Rehabil.* **2003**, *25*, 120–126. [CrossRef] [PubMed]
12. Youdas, J.W.; Carey, J.R.; Garrett, T.R. Reliability of measurements of cervical spine range of motion–comparison of three methods. *Phys. Ther.* **1991**, *71*, 98–104; discussion 105–106. [CrossRef] [PubMed]
13. Timurtaş, E.; Avcı, E.E.; Mate, K.; Karabacak, N.; Polat, M.G.; Demirbüken, İ. A mobile application tool for standing posture analysis: Development, validity, and reliability. *Ir. J. Med. Sci.* **2021**, *191*, 2123–2131. [CrossRef] [PubMed]

14. Szucs, K.A.; Brown, E.V.D. Rater reliability and construct validity of a mobile application for posture analysis. *J. Phys. Ther. Sci.* **2018**, *30*, 31–36. [CrossRef] [PubMed]

15. Moreira, R.; Fialho, R.; Teles, A.S.; Bordalo, V.; Vasconcelos, S.S.; Gouveia, G.P.M.; Bastos, V.H.; Teixeira, S. A computer vision-based mobile tool for assessing human posture: A validation study. *Comput. Methods Programs Biomed.* **2022**, *214*, 106565. [CrossRef]

16. Heyward, V.H. Advanced fitness assessment and exercise prescription. *Med. Sci. Sport. Exerc.* **1992**, *24*, 278. [CrossRef]

17. Koo, T.K.; Li, M.Y. A Guideline of Selecting and Reporting Intraclass Correlation Coefficients for Reliability Research. *J. Chiropr. Med.* **2016**, *15*, 155–163. [CrossRef]

18. Belli, G.; Toselli, S.; Latessa, P.M.; Mauro, M. Evaluation of Self-Perceived Body Image in Adolescents with Mild Idiopathic Scoliosis. *Eur. J. Investig. Health Psychol. Educ.* **2022**, *12*, 319–333. [CrossRef]

19. Boguszewski, D.V.; Cheung, E.C.; Joshi, N.B.; Markolf, K.L.; McAllister, D.R. Male-Female Differences in Knee Laxity and Stiffness: A Cadaveric Study. *Am. J. Sports Med.* **2015**, *43*, 2982–2987. [CrossRef]

20. Raine, S.; Twomey, L.T. Head and shoulder posture variations in 160 asymptomatic women and men. *Arch. Phys. Med. Rehabil.* **1997**, *78*, 1215–1223. [CrossRef]

21. Iacob, S.M.P.; Chisnoiu, A.M.P.; Lascu, L.M.P.; Berar, A.M.P.; Studnicska, D.M.; Fluerasu, M.I.P. Is PostureScreen®Mobile app an accurate tool for dentists to evaluate the correlation between malocclusion and posture? *Cranio* **2020**, *38*, 233–239. [CrossRef]

22. Zhao, X.; Tsujimoto, T.; Kim, B.; Katayama, Y.; Tanaka, K. Characteristics of foot morphology and their relationship to gender, age, body mass index and bilateral asymmetry in Japanese adults. *J. Back Musculoskelet Rehabil.* **2017**, *30*, 527–535. [CrossRef] [PubMed]

23. Redmond, A.C.; Crane, Y.Z.; Menz, H.B. Normative values for the Foot Posture Index. *J. Foot Ankle Res.* **2008**, *1*, 6. [CrossRef] [PubMed]

24. Gimunová, M.; Válková, H.; Kalina, T.; Vodička, T. The relationship between body composition and foot posture index in Special Olympics athletes. *Acta Bioeng. Biomech.* **2019**, *21*, 47–52. [CrossRef] [PubMed]

Journal of
Functional Morphology and Kinesiology

MDPI

Article

Relation between Photogrammetry and Spinal Mouse for Sagittal Imbalance Assessment in Adolescents with Thoracic Kyphosis

Guido Belli [1], Stefania Toselli [1], Mario Mauro [1,*], Pasqualino Maietta Latessa [1] and Luca Russo [2]

[1] Department of Sciences for Life Quality Studies, University of Bologna, 47921 Rimini, Italy; guido.belli@unibo.it (G.B.)

[2] Department of Human Sciences, IUL Telematic University, 50122 Florence, Italy; l.russo@iuline.it

*   Correspondence: mario.mauro4@unibo.it

Abstract: The evaluation of postural alignment in childhood and adolescence is fundamental for sports, health, and daily life activities. Spinal Mouse (SM) and photogrammetry (PG) are two of the most debated tools in postural evaluation because choosing the proper instrument is also important to avoid false or misleading data. This research aims to find out the best linear regression models that could relate the analytic kyphosis measurements of the SM with one or more PG parameters of body posture in adolescents with kyphotic posture. Thirty-four adolescents with structural and non-structural kyphosis were analyzed (13.1 ± 1.8 years; 1.59 ± 0.13 m; 47.0 ± 12.2 kg) using SM and PG on the sagittal plane in a standing and forward-bending position, allowing us to measure body vertical inclination, trunk flexion, and sacral inclination and hip position during bending. The stepwise backward procedure was assessed to estimate the variability of the grade of inclination of the spine and thoracic spine curvature with fixed upper and lower limits, evaluated with SM during flexion. In both models, the PG angle between the horizontal line and a line connecting the sacral endplate–C7 spinous process and the PG hip position were the best regressors (*adjusted-R^2* SM bend = 0.804, $p < 0.001$; *adjusted-R^2* SM fixed bending = 0.488, $p < 0.001$). Several Spinal Mouse and photogrammetry parameters showed significant correlations, especially when the Spinal Mouse measurements were taken when the adolescents were in the forward-bending position. Physicians and kinesiologists may consider photogrammetry as a good method for spinal curve prediction.

Keywords: kyphosis; spinal mouse; photogrammetry; kinesiology; postural evaluation

Citation: Belli, G.; Toselli, S.; Mauro, M.; Maietta Latessa, P.; Russo, L. Relation between Photogrammetry and Spinal Mouse for Sagittal Imbalance Assessment in Adolescents with Thoracic Kyphosis. *J. Funct. Morphol. Kinesiol.* **2023**, *8*, 68. https://doi.org/10.3390/jfmk8020068

Academic Editor: Luca Petrigna

Received: 31 March 2023
Revised: 13 May 2023
Accepted: 17 May 2023
Published: 19 May 2023

1. Introduction

Postural alignment in childhood and adolescence can be considered one of the most important sources of worry for parents, with particular attention being paid to the spine. Altered postural alignments can be classified as structural and non-structural misalignments, even if the postural appearance of these disturbances may be similar [1]. Structural misalignments indicate the presence of morphological abnormalities within the bones and soft tissues. Conversely, non-structural misalignments do not show any bone disorder but evidence a non-anatomic spine alignment with a moderate-to-good degree of self-correction [2]. Both structural and non-structural misalignments can affect the sagittal balance of the spine [3–5].

With specific regard to the sagittal plane, several classifications have been previously reported [2,6–9]. Within these classifications, kyphotic posture is frequently and easily recognized by parents. The thoracic spine in children and adolescents can be defined in a normal range as being between 20–40° degrees [1], and hyperkyphosis diagnosis is considered beyond 45° [10,11]. Nevertheless, other authors suggest that the average values of thoracic kyphosis are $42.0° \pm 10.6°$ and $45.8° \pm 10.4°$ from a cohort of 167 children of 8.1 ± 2.0 years and 479 adolescents of 13.6 ± 1.9 years, respectively [12].

Measuring procedures in posture can be widely utilized, such as with posturography, to analyze the effect of specific interferences on postural control [13–15]; or they can be analytical and descriptive, such as with photogrammetric analysis of the spine, with a focus on sagittal balance assessment using non-radiographic methods as an example [16]. In the last 15 years, the main three non-radiographic methods for spine evaluation were rasterstereography, skin-surface mouse, and photogrammetry [17–20]. Rasterstereography is mainly represented by two kinds of measuring methods: (1) the first one uses the analysis of the light projected on the subject's skin; it is reliable and represents the most widespread solution for the application of rasterstereography [21–23]; (2) the second one uses an infrared and time-of-flight 3D RGB camera, and it seems to be reliable as well [24]. The skin-surface mouse is mainly represented by Spinal Mouse® (IDIAG, Fehraltorf, Switzerland), a valid and reliable tool for spine assessment, in particular for kyphotic posture [25,26], that can be rolled along the profile of the spine measuring the vertebral shape and angulation [16,27–29]. Finally, photogrammetry, in particular 2D modality, is one of the most used and cheapest tools for kinematic and geometrical analysis of motion and even posture. It can be performed using different software, and many of them are valid and reliable [30–32]. All these non-radiographic methods evidence some advantages but show several limits for spine evaluation. Researchers, but even more so professionals, should always balance the cost/effectiveness ratio as well as the ease and accuracy of the measurement. Rasterstereography usually offers a wide range of postural parameters; it is very fast to use but represents a high-cost tool. Spinal Mouse has a lower cost, but the price range is not accessible for all; the accuracy is high, as is the software analysis, but this is only focused on the spine [25,27]. Finally, photogrammetry is the cheaper one, as well as the less "smart". Depending on software features and user skills, photogrammetry can be more or less "user friendly"; at the same time, it also allows the user to obtain measures of the whole body, not only of the spine [32].

Within this scenario, it could be very interesting for professionals to use photogrammetry, the easiest and cheaper way, as a global screening tool to detect in advance the signs of a spine misalignment, with particular regard to the kyphosis curvature. Considering the whole body as a kinetic chain [33], and that photogrammetry could easily measure the alignment of the whole body with good validity and reliability [32], a relation could be hypothesized between thoracic spine behavior and one or more photogrammetric measurements of body posture. Although radiographic and photogrammetric procedures have been previously investigated in order to quantify thoracic kyphosis and lumbar lordosis [34,35], the comparison between Spinal Mouse® evaluation and photogrammetry is lacking. Since Spinal Mouse® can be used in several body positions (upright standing, forward trunk bending, seated side-bending, for example), additional information could be obtained from this device. Therefore, this study aims to find out a model of regression that could relate global and analytic measurements of the Spinal Mouse® on the sagittal plane with one or more photogrammetric parameters of body posture in adolescents with kyphotic posture.

2. Materials and Methods

2.1. Design and Participants

This is a cross-sectional study design. Participants were recruited from Fisiokinè Medical Centre (Scandiano, Reggio Emilia, Italy). The criteria of selection included a diagnosis of increased thoracic kyphosis (postural or structural hyperkyphosis), no history of musculoskeletal or neurological pain in the last 3 months, no prior surgical intervention for spine disorders, and aged between 10 to 16 years old. No gender restrictions were defined. All participants were informed and gave voluntary consent to participate in the study. Parents' consent was requested, since participants were younger than 18 years old. The privacy criteria were met. The study was approved by the Bioethics Committee of the University of Bologna and was conducted in accordance with the guidelines of the Declaration of Helsinki; the project identification code was n.2.18.

During the recruitment phase, each participant completed the anamnesis investigation. All specific medical reports were collected and analyzed to meet the selection criteria. The enrolment phase lasted 6 months, from June to December 2022.

2.2. Measurements Instruments

2.2.1. Spine Analysis

To evaluate spinal curves and trunk alignment, the SpinalMouse® (IDIAG M360®, Mülistrasse 18, CH-8320 Fehraltorf, Switzerland) device was used. It is a non-invasive computer-assisted medical device that quantifies the curvature and mobility of the spinal column in the frontal and sagittal planes by gliding manually along the spine [28,36]. Data are sampled every 1.3 mm while the mouse is rolled from vertebra C7 to S3, giving a sampling frequency of approximately 150 Hz. Results are wirelessly transferred to a computer, where the IDIAG software displays vertebral positions, joint angles, and spinal measurements. A recent study reported a high correlation between Cobb angle evaluated with X-ray and intra (ICC = 0.872) or inter-observer (ICC = 0.962) SpinalMouse® measurements on the frontal plane [37]. In addition, this device evidenced excellent intra-rater reliability for the analysis of sagittal thoracic and lumbar curvature and mobility in hyperkyphosis [25]. In the present study, the SpinalMouse® measurements were performed by a trained specialist with more than five years of experience. Data were collected in a quiet and well-lit environment with a comfortable temperature [38,39]. The evaluation was settled in the morning to avoid positional differences in the spine due to fatigue and/or daily stress factors. After undressing the upper body, the C7–S3 vertebral spinal processes were determined and marked with a dermographic pen by the specialist while the patient was standing up in the anatomical position. Measurements were performed in 3 different trunk positions during standing: neutral, maximal flexion, and extension (sagittal plane evaluation). In the neutral position, the participant was asked to maintain a relaxed position, looking and facing horizontally toward the wall, with the feet shoulder-width apart and with straight knees and arms by the side. In maximal flexion, the subject was asked to flex the trunk with extended legs as far as possible, aiming to touch the ground with fingertips. In maximal extension, the participant was asked to cross their arms in front of the chest and extend the trunk as far as possible, without extension of the cervical spine. SpinalMouse® was then moved downwards along the spinal criteria points, in each position. Participants did not perform a warm-up before the examination. Some specific measures were extracted and analyzed from all raw data available. The eight variables were: the inclination of the spine in standing (SM stand); the inclination of the spine during flexion (SM bend); thoracic spine curvature with fixed upper and lower limits (first and last thoracic vertebra) in standing (SM fixed stand) and during flexion (SM fixed bend); thoracic spine curvature with physiological upper and lower limits (defined by Spinal Mouse software) in standing (SM phys. stand) and during flexion (SM phys. bend); and spine length in standing position (SM Rachid stand) and during flexion (SM Rachid bend). Figure 1 shows some of all Spinal Mouse possible measurements displayed using IDIAG M360 software and used in the present study.

2.2.2. Photogrammetric Postural Analysis

Postural evaluation using photogrammetry has been previously demonstrated to be a reliable method in young people with postural misalignments [40,41]. Recently, photogrammetric measurements of thoracic kyphosis showed excellent test–retest reliability (ICC = 0.97; SEM = 1.67; MDC = 4.62) in adolescents with hyperkyphosis and evidenced a strong correlation between the values obtained with this technique and radiography methods [42]. In the present study, 2 digital photographs (standing right-side and standing with trunk flexion) were recorded using a portable device (Tablet Huawei® Mediapad, Huawei Base, Bantian, Longgang District, Shenzhen, China) to analyze the sagittal plane. The device was set on a tripod, three meters away from the line marking the position of the participant. The height of the tripod was adjusted so the middle of the objective lens was

100 cm above the ground [43]. Each participant was initially positioned in front of the camera with a postural grid (ATS®, Largo Cairoli 10, 52100 Arezzo, Italy) on the back, then made to turn their body to the left to show the right side perpendicular to the lens, with feet placed in a fixed position over a specific area (standing right-side position —Figure 2A). Successively, participants flexed the trunk and remained in the forward-bending position (standing trunk flexion position—Figure 2B–D). The APECS-AI Posture Evaluation and Correction System® (New Body Technology SAS, 12 Rue Pierre Semard, Incubagem 38000 Grenoble, France) was used to evaluate absolute and relative angles in the sagittal plane [29,32]. Specifically, the following angles were investigated: body vertical inclination (absolute angle between the vertical line and a line connecting the lateral malleolus–tragus of the ear: PG mall–tragus); trunk flexion (absolute angle between the horizontal line and a line connecting sacral endplate—C7 spinous process, PG trunk bend); sacral inclination during bending (absolute angle between horizontal line and a tangent line to the sacral dorsum, PG sacrum bend); hip position during bending (absolute angle between the vertical line and a line connecting the lateral malleolus–greater trochanter, PG hip bend). To better detect previous anatomic landmarks during photographic analysis, an adhesive tape was applied to the skin [43].

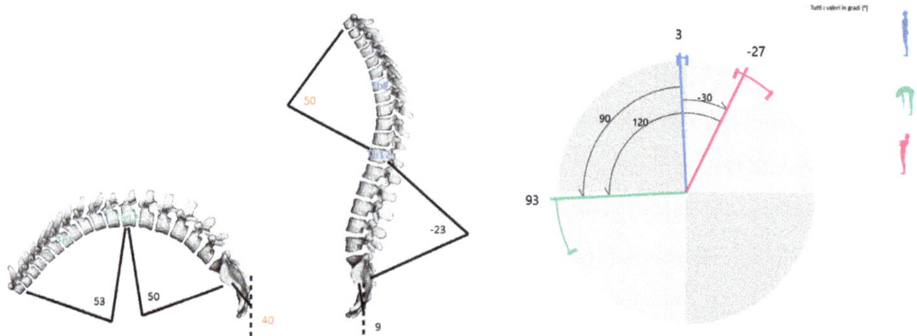

Figure 1. Example of Spinal Mouse report—thoracic and lumbar curvature (**left side**); body inclination (**right side**).

Figure 2 shows the four angles analyzed with the APECS application.

2.3. Statistical Procedure

The descriptive statistics were reported as the mean, standard deviation (std), minimum (min), and maximum (max) values for each variable. The variables' distribution was verified with the Shapiro–Wilk test. The Pearson product–moment (r) was calculated to measure the degree of correlation between the variables. To perform the best regression model, the stepwise backward procedure was assessed with a significant level for entry or removal to or from the model equal to 0.10. The model's heteroskedasticity was checked using the Breusch–Pagan/Cook–Weisberg test. The multicollinearity was checked using the variance inflation factor (VIF), and a value lower than 5 was considered acceptable (moderate correlation) [44]. The Cook's distance plot was performed to look for the outlier presence, with a threshold settled at n/4. If one or more outliers affected the model, they were removed, and a new model was performed. The adjusted R^2 was calculated to report the goodness-of-fit for the proposed model. Additionally, the F value, the root MSE, the regression coefficient (β), the standard error, the student's *t*-test value, and the 95% confidence interval were reported. The significance level was settled at ≤ 0.05. Finally, the Bland–Altman plot, the pairwise correlation, and the concordance correlation coefficient (CCC) were computed to compute the degree of agreement between the Spinal Mouse and photogrammetry [45].

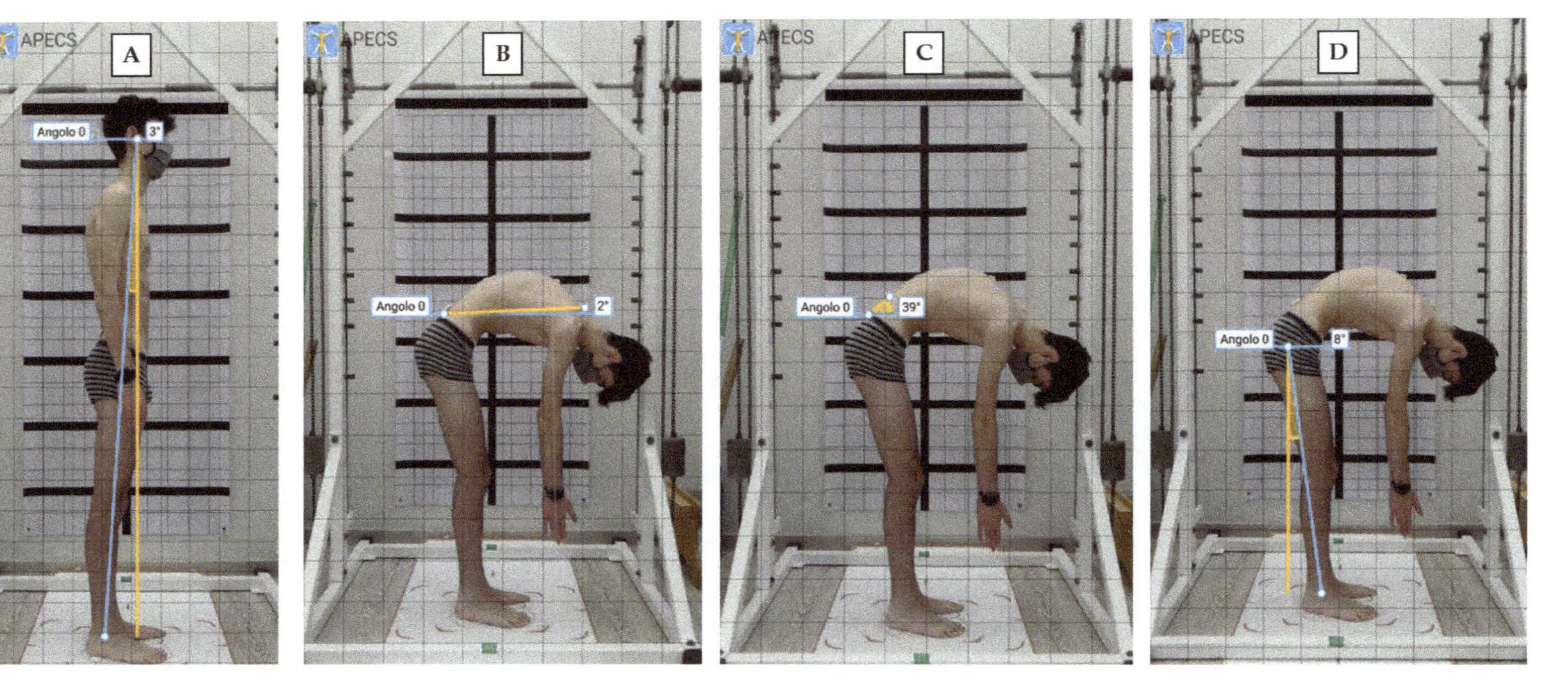

Figure 2. Photogrammetric analysis—(**A**) body vertical inclination (PG mall–tragus); (**B**) trunk flexion (PG trunk bend); (**C**) sacral inclination (PG sacrum bend); (**D**) hip position (PG hip bend).

3. Results

Table 1 shows the descriptive statistics for all the variables. No missing value was met for all variables (n = 34). Similar values were found between fixed and physiological Spinal Mouse evaluation, in both standing (45.94 ± 8.24 and 48.29 ± 9.14) and bending (63.71 ± 10.23 and 63.41 ± 9.45) positions. The highest grade was found in the inclination of the spine during flexion (122°), whereas the smallest value was the grade of the inclination of the spine in a standing position (−6°).

Table 1. Descriptive statistics.

Variable (n = 34)	Mean	Std	Min	Max
Age [year]	13.06	1.84	10	18
Height [cm]	158.56	12.73	139	189
Weight [kg]	47.03	12.19	26	75
PG mall-tragus [°]	2.79	1.20	1	6
PG trunk bend [°]	91.71	13.35	68	121
PG sacrum bend [°]	37.15	12.02	4	55
PG hip bend [°]	8.71	2.69	3	15
SM stand [°]	1.68	2.81	−6	7
SM bend [°]	90.56	13.16	65	122
SM fixed stand [°]	45.94	8.24	29	62
SM phys. stand [°]	48.29	9.14	30	69
SM fixed bend [°]	63.71	10.23	45	87
SM phys. bend [°]	63.41	9.45	46	82
SM Rachid stand [mm]	457.18	44.02	394	593
SM Rachid bend [mm]	538.44	51.33	449	673

Note: n, number of observations; std, standard deviation; min, minimum value observed; max, maximum value observed; PG, photogrammetry; mall, malleolus; bend, bending; SM, Spinal Mouse; stand, standing; phys., physiological kyphosis.

Table 2 shows the variables' correlation matrix. Generally, high Pearson correlation coefficients between Spinal Mouse and photogrammetry measurements were found in the bending position ($p < 0.05$). In particular, a wide positive correlation appeared among the inclination of the spine during trunk flexion and trunk flexion, calculated as the absolute angle between the horizontal line and a line connecting the sacral endplate–C7 spinous process (r = 0.839, $p < 0.001$). Differently, the grade of the inclination of the spine during flexion decreased with the increasing of the absolute angle between a horizontal line and a tangent line to the sacral dorsum (r = −0.732, $p < 0.001$). Additionally, the thoracic spine curvature with fixed upper and lower limits in the bending position showed high correlations with the trunk flexion and the sacral inclination detected using photogrammetry ($p < 0.001$).

Table 2. Variables' correlation matrix.

	PG Mall–Tragus	PG Trunk Bend	PG Sacrum Bend	PG Hip Bend	SM Stand	SM Bend	SM Fixed Stand	SM Phys. Stand	SM Fixed Bend	SM Phys. Bend	SM Rachid Stand	SM Rachid Bend
PG mall–tragus	-											
PG trunk bend	0.003	-										
PG sacrum bend	−0.076	−0.860 *	-									
PG hip bend	−0.176	−0.479 *	0.327	-								
SM stand	−0.045	−0.059	0.061	0.379 *	-							
SM bend	−0.041	0.839 *	−0.732 *	−0.237	0.119	-						
SM fixed stand	0.013	−0.204	0.330	−0.151	−0.098	−0.29	-					
SM phys. stand	0.036	−0.024	0.195	−0.182	−0.020	−0.17	0.908 *	-				
SM fixed bend	0.138	−0.526 *	0.559 *	−0.063	−0.076	−0.61 *	0.403 *	0.251	-			

Table 2. *Cont.*

	PG Mall–Tragus	PG Trunk Bend	PG Sacrum Bend	PG Hip Bend	SM Stand	SM Bend	SM Fixed Stand	SM Phys. Stand	SM Fixed Bend	SM Phys. Bend	SM Rachid Stand	SM Rachid Bend
SM phys. bend	0.060	−0.316	0.350 *	0.024	−0.149	−0.41 *	0.404 *	0.420 *	0.637 *	-		
SM Rachid stand	0.392 *	0.000	0.042	−0.222	−0.102	−0.08	0.206	0.319	0.121	0.272	-	
SM Rachid bend	0.273	−0.014	0.065	−0.243	−0.258	−0.1	0.180	0.243	0.228	0.332	0.914 *	-

Note: PG, photogrammetry; mall, malleolus; bend, bending; SM, Spinal Mouse; stand, standing; phys., physiological kyphosis; * *p*-value $\leq$ 0.05.

Linear Regression Models

Table 3 shows the result of the stepwise procedure. Two outliers were removed from the first model, and the two photogrammetry measurements in bending, such as the angle between the horizontal line and a line connecting the sacral endplate–C7 spinous process and the hip position, explained 80.4% of the variability of the spine inclination during trunk flexion measured with the Spinal Mouse on 32 adolescents. The Breusch-Pagan/Cook-Weisberg test accepts the null hypothesis of heteroskedasticity absence ($\chi^2_{(1)}$ = 0.77, *p* = 0.711). The mean VIF was 1.19.

Table 3. Best linear regression model for bending Spinal Mouse estimation.

Source	SS	df	MS	n	$F_{(2,\,28)}$	*p*	R^2	Adj. R^2	Root MSE
Model	4513.86	2	2256.93	32	64.44	<0.001	0.816	0.804	5.918
Residual	1015.64	29	35.02						
Total	5529.5	31	178.37						
SM bending		β	SE	*t*	*p*		95% CI		
PG hip bend		0.86413	0.4394	1.97	0.059	−0.0346	1.763		
PG trunk bend		1.0557	0.0958	11.02	<0.001	0.86	1.252		
Intercept		−12.749	10.874	−1.17	0.251	−34.99	9.491		

Note: SS, squared sums; df, degrees of freedom; MS, squared means; n, number of observations; *F*, Snedecor–Fisher's test; *p*, *p*-value; R^2, the goodness-of-fit; Adj., adjusted; MSE, mean of squares error; β, regression coefficient; *t*, Student's test; CI, confidence interval.

Figure 3 shows the scatterplots with the errors of bending SM and each regressor, respectively.

Figure 4 shows the Bland–Altman graph (A) and scatterplot with Spinal Mouse and the new model (B). The concordance correlation coefficient was 0.995 and Pearson's r = 0.904 (mean = 90.125 $\pm$ 12.067).

The new equation to estimate the Spinal Mouse degree in bending is

$$SM\ bend = (0.86413 \cdot PG\ hip\ bend) + (1.0557 \cdot PG\ trunk\ bend) - 12.749$$

Table 4 shows the result of the stepwise procedure on the thoracic spine curvature with fixed upper and lower limits, in the bending position. Three outliers were removed from the first model, and the two photogrammetry measurements in bending explained 48.79% of the Spinal Mouse fixed kyphosis variability in 31 adolescents. The Breusch–Pagan/Cook–Weisberg test accepts the null hypothesis of heteroskedasticity absence ($\chi^2_{(1)}$ = 1.80, *p* = 0.179). The mean VIF was 1.32.

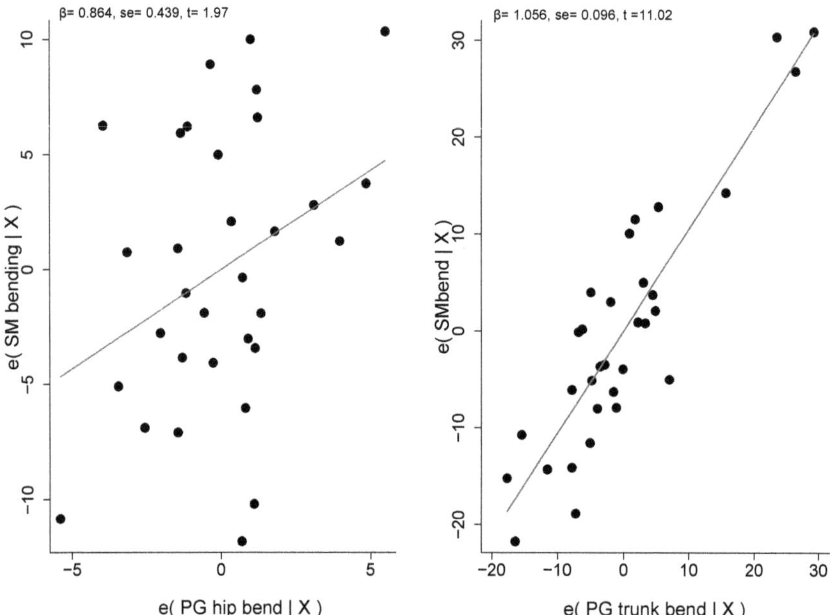

Figure 3. Errors' scatterplots with SM and PG model regressors. Note: β, regressor coefficient; se, standard error; t, Student's *t*-test.

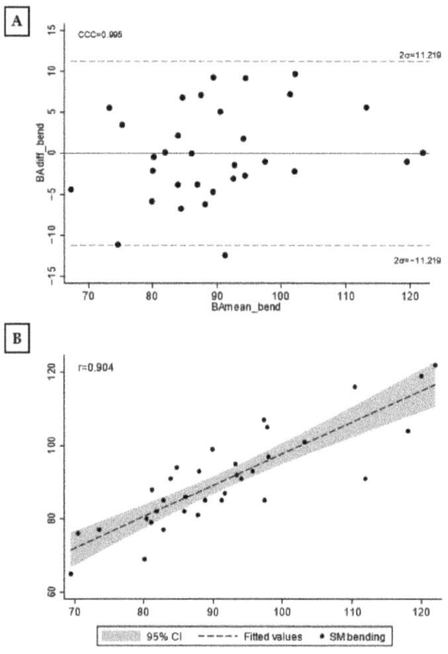

Figure 4. Bland–Altman plot (**A**) and scatterplot (**B**) with gold standard and new estimated model. Note: CCC, concordance correlation coefficient; σ, standard deviation; r, Pearson correlation coefficient; CI, confidence interval.

Table 4. Best linear regression model for Spinal Mouse fixed kyphosis estimation.

Source	SS	df	MS	n	$F_{(2,\,28)}$	p	R^2	Adj. R^2	Root MSE
Model	1701.8126	2	850.90628	31	15.29	<0.001	0.522	0.488	7.46
Residual	1558.0584	28	55.644943						
Total	3259.871	30	108.66237						
SM fixed bending	β		SE	t	p	[95% conf.	interval]		
PG hip bend	−2.122705		0.6704444	−3.17	0.004	−3.496048	−0.749361		
PG trunk bend	−0.678946		0.1232905	−5.51	<0.001	−0.931495	−0.426397		
Intercept	144.2784		15.00178	9.62	<0.001	113.5486	175.0081		

Note: SS, squared sums; df, degrees of freedom; MS, squared means; n, number of observations; F, Snedecor–Fisher's test; p, p-value; R^2, goodness-of-fit; Adj., adjusted; MSE, mean of squares error; β, regression coefficient; t, student's test; CI, confidence interval.

Figure 5 shows the scatterplots with the errors of SM fixed kyphosis measured in bending position and each regressor, respectively.

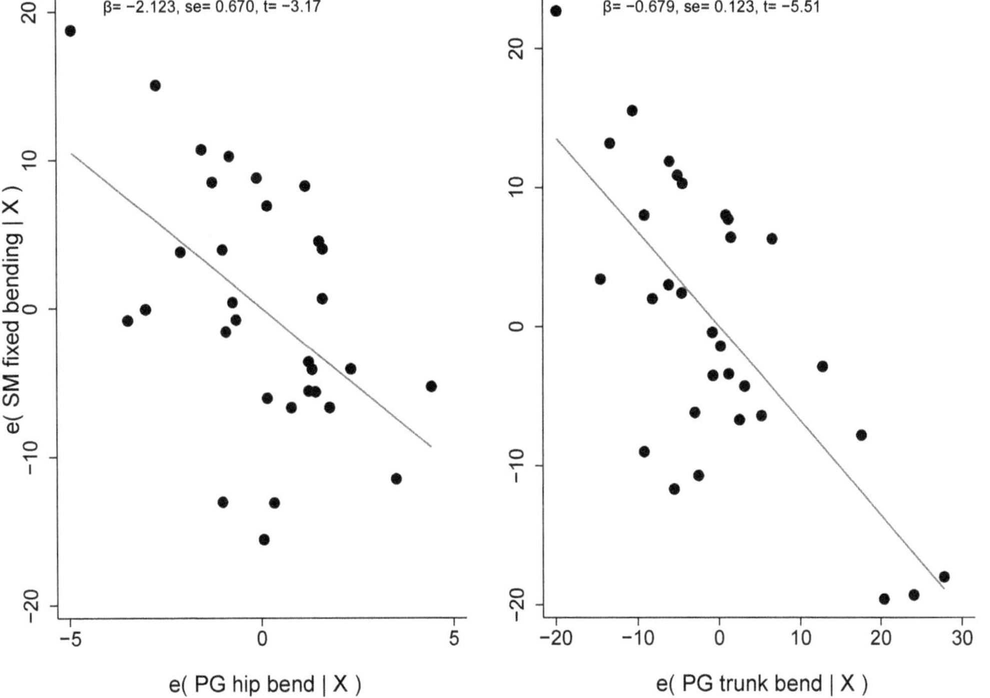

Figure 5. Errors' scatterplots with SM fixed kyphosis and PG model regressors. Note: β, regressor coefficient; se, standard error; t, Student's t-test.

Figure 6 shows the Bland–Altman graph (A) and scatterplot with Spinal Mouse fixed kyphosis and the new model (B). The concordance correlation coefficient was 0.686 and Pearson's r = 0.723.

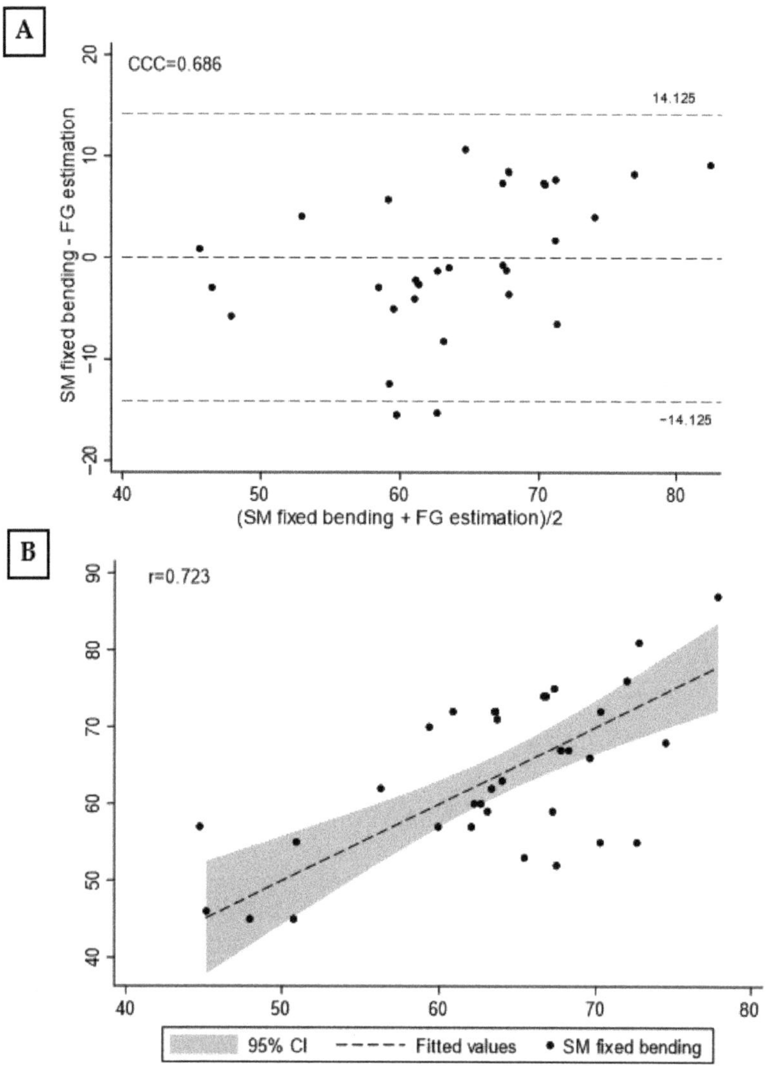

Figure 6. Bland–Altman plot (**A**) and scatterplot (**B**) with fixed Spinal Mouse in bending position and the new estimated model. Note: CCC, concordance correlation coefficient; σ, standard deviation; r, Pearson correlation coefficient; CI, confidence interval.

The new equation to estimate the Spinal Mouse fixed kyphosis degree in the bending position is

$$SM\ fixed\ bending = (-2.122705 \cdot PG\ hip\ bend) + (-0.678946 \cdot PG\ trunk\ bend) + 144.2784$$

4. Discussion

The present study aimed to correlate photogrammetry and Spinal Mouse® during the postural evaluation of adolescents with a diagnosis of structural or non-structural hyperkyphosis. Current findings evidence a positive correlation between some measurements of standing trunk flexion performed with both devices and highlight how photogrammetry could explain 80.4% of Spinal Mouse variability during forward bending. To the best of our knowledge, this is the first study that compared SM and PG.

In recent years, several authors analyzed the reliability and validity of non-radiographic methods during sagittal balance assessment in different populations. Since an adequate anterior–posterior balance condition is fundamental to maintaining an upright, efficient, and painless posture, the evaluation of the sagittal profile has gained much relevance in spinal pathology [46]. In this direction, Cohen et al. [16] reported that plumbline, surface topography, infrared motion analysis, and SM show moderate-to-high validity and reliability. In their systematic review of 14 articles, authors suggested that these methods can be a non-invasive approach to monitor global sagittal balance, even if specific limitations are present and spinopelvic parameters represent the "gold standard" (sacral vertical axis, pelvic tilt, and sacral slope, as an example). Furthermore, Barret et al. [26] evidenced that SM, the Debrunner kyphometer, and the Flexicurve index have the strongest levels of reliability among 15 non-radiographic methods to assess thoracic kyphosis. Starting from this consideration and previous research [25,47,48], SM was included in our study. Roghani et al. evaluated SM reliability on a sample of women with and without hyperkyphosis (aged between 60–80 years), with a focus on thoracic and lumbar curvature, pelvic position, trunk inclination, and spine mobility. The results evidenced a high intra-rater reliability for all measurements in both groups (ICC: 0.89–0.99), with standard error of the means (SEMs) ranging from $1.02°$ to $2.06°$ and from $1.15°$ to $2.22°$ in the hyperkyphosis and normal group, respectively. In addition, the minimal detectable change (MDC) ranged from $2.85°$ to $5.73°$ in the hyperkyphosis group and from $3.20°$ to $6.17°$ in the normal group. The authors concluded that SM is a useful, easy, and low-risk device to assess spinal curvature and mobility. Demir et al. evaluated SM test–retest reliability in 28 female adolescents (aged between 15–18 years) during upright standing on the frontal and sagittal planes. Their results evidenced good reliability for thoracic and lumbar curvature on the sagittal plane and confirmed the use of these tools for "in-field" screening and clinical assessment. Muyor and collaborators analyzed the Sit-and-Reach Test and Toe-Touch test (same as the forward bending test) using SM to define the criterion validity of both tests during hamstring flexibility assessment. The study involved 141 athletes from different sports (tennis players, kayakers, canoeists, cyclists), all aged between 15–17 years. Research findings suggested that pelvic tilt and lumbar motion have a greater impact on test scores than hamstring flexibility (measured with a passive straight leg test). In the present study, only some SM measurements were investigated. Specifically, fixed and physiologic thoracic curvature, body inclination, and spine length in upright and bending positions were chosen. The reason is mainly related to: (1) the sample features (adolescents with postural or structural hyperkyphosis); (2) to compare these parameters with some specific photogrammetric variables; (3) to attempt to find a few quick and easy-to-detect photogrammetric landmarks [31,49]. Mean values for SM thoracic curvature in a standing position were $45.9°$ and $48.2°$ for fixed and physiological kyphosis, respectively. These values are slightly above the reported range for normal curvature ($20–40°$) and evidence of slight hyperkyphosis. Similar values have been reported in different sports players performing in flexed positions (kayak, canoa, tennis) and that were aged between 15–17 [50], as well as for adolescents aged between 12–15 [51]. Anyway, it must be considered that higher values have been found for structural spine misalignment diagnosis (range of $50–62°$) compared to non-structural (range of $38–50°$).

For photogrammetric analysis, the application APECS-AI Posture Evaluation and Correction System® has been used. This tool is an easy and low-cost program that allows one to assess body posture in different positions. Recently, Trovato et al. [32] evaluated a sample of 50 males and 50 females (mean age 23.4 years) to investigate gender differences in anterior coronal, posterior coronal, and sagittal planes. Their results evidenced good reproducibility for most of the 24 variables analyzed and reported some gender-related features. In the present study, only four sagittal parameters were investigated to assess the anterior–posterior balance in upright and bending positions. This selection has been defined for the abovementioned reasons. In particular, trunk inclination in both positions and pelvic motion during forward bending have been correlated with thoracic curvature. As regards body alignment, the results evidenced a mean value of 2.7° and 91.7° during standing or bending positions, respectively. Sagittal vertical inclination (upright posture) is similar to the reference value of 2.6° highlighted by Trovato et al. in their sample and superior to the reference value of 1.73° defined by Krawky et al. [31]. The difference could be related to the photogrammetry techniques reported in both studies [43,52]. Results did not show a correlation between PG and SM in relation to this parameter (2.6° vs. 1.6°, respectively). Since the body vertical alignment using PG was calculated by using the absolute angle between the vertical line and a line connecting the lateral malleolus–tragus of the ear, while SM evaluated it from the C7–S3 connection and vertical line, this result could be expected. Conversely, a significant positive correlation was found between the two devices during the bending position. In this postural assessment, PG and SM used a similar technique (absolute angle between the horizontal line and a line connecting the sacral endplate–C7 spinous process for PG or absolute angle between the vertical line and a line connecting the S3–C7 spinous process for SM). The landmarks for pelvic parameters during forward bending were defined as reported by Carregaro et al. [53]. Since adolescents with hyperkyphosis show a low level of flexibility across the muscular posterior chain, sacral inclination and hip position were calculated to investigate pelvic displacement in the sagittal plane [6,33,48]. PG Sacrum inclination evidenced a significant correlation with SM trunk inclination during bending, and linear regression reported that trunk bending added to hip bending explained 80.4% of SM variability in relation to this parameter (first model). Furthermore, the same PG variables explained 48.7% of SM variability concerning thoracic curvature during bending. These results highlight how the photogrammetric analysis of the pelvic region is deeply connected with spine inclination in this kind of subject.

Since the present study aims to correlate quick and easy-to-detect PG measurements with Spinal Mouse®, the abovementioned variable was chosen. Previous researchers analyzed different pelvic parameters using photogrammetry (pelvic horizontal alignment and hip angle, for example), especially in the upright position [31,49]; however, there is a lack of research on the bending posture [53]. Specifically, the focus has been more addressed on frontal plane analysis than on the sagittal plane [54]. Finally, SM and PG have recently been investigated in body perception in adolescents with idiopathic scoliosis [29].

Limitations

The sample size, participants' ages, morpho-structural characteristics, differences in spine misalignment, and level of physical activity represent limits in our work. Furthermore, Spinal Mouse® and photogrammetry could be affected from measurement errors. Although the reliability and validity of both methods have been previously described, the occurrence of possible evaluation errors must be considered (markers positioning and angles investigation using photogrammetry or skin surface contact during Spinal Mouse® analysis, for example). Future investigations are needed investigating the role of Spinal Mouse® and photogrammetry during postural evaluation in adolescents with structural and non-structural hyperkyphosis and in relation to specific phases of treatment.

5. Conclusions

The present study suggests that photogrammetry can be considered an easy, inexpensive, and rapid tool for postural screening in adolescents with both structural and non-structural hyperkyphosis. Photogrammetry can be a useful alternative in the absence of other specific instruments, although it should be noted that photogrammetry is more suitable for screening rather than diagnosis. Photogrammetry is significantly correlated with the SM parameters when the analysis is taken during spine forward-bending. Specifically, the pelvic motion measured with photogrammetry can predict over 80% of the SM spine inclination. Differently, these instruments exhibited low correlations in standing positions. In conclusion, Kinesiologists and professionals involved in postural assessment are encouraged to use photogrammetry in bending as a first-line assessment tool to evaluate the sagittal spinal profile of adolescents.

Author Contributions: Conceptualization, G.B. and L.R.; methodology, G.B., L.R. and M.M.; software, P.M.L.; validation, G.B., L.R. and M.M.; formal analysis, M.M. and S.T.; investigation, G.B.; resources, P.M.L. and S.T.; data curation, G.B., L.R. and M.M.; writing—original draft preparation, G.B., L.R. and M.M.; writing—review and editing, G.B. and L.R.; visualization, S.T.; supervision, P.M.L.; project administration, G.B. All authors have read and agreed to the published version of the manuscript.

Funding: This research received no external funding.

Institutional Review Board Statement: The study was conducted in accordance with the Declaration of Helsinki and approved by the Institutional Ethics Committee of University of Bologna (n. 2.18).

Informed Consent Statement: Informed consent was obtained from all subjects involved in the study.

Data Availability Statement: The data presented in this study are available on request from the corresponding author.

Conflicts of Interest: The authors declare no conflict of interest.

References

1. Feng, Q.; Wang, M.; Zhang, Y.; Zhou, Y. The Effect of a Corrective Functional Exercise Program on Postural Thoracic Kyphosis in Teenagers: A Randomized Controlled Trial. *Clin. Rehabil.* **2018**, *32*, 48–56. [CrossRef] [PubMed]
2. Czaprowski, D.; Stoliński, Ł.; Tyrakowski, M.; Kozinoga, M.; Kotwicki, T. Non-Structural Misalignments of Body Posture in the Sagittal Plane. *Scoliosis* **2018**, *13*, 6. [CrossRef] [PubMed]
3. Le Huec, J.C.; Thompson, W.; Mohsinaly, Y.; Barrey, C.; Faundez, A. Sagittal Balance of the Spine. *Eur. Spine J.* **2019**, *28*, 1889–1905. [CrossRef] [PubMed]
4. Abelin-Genevois, K. Sagittal Balance of the Spine. *Orthop. Traumatol. Surg. Res.* **2021**, *107*, 102769. [CrossRef]
5. Roussouly, P.; Nnadi, C. Sagittal Plane Deformity: An Overview of Interpretation and Management. *Eur. Spine J.* **2010**, *19*, 1824–1836. [CrossRef]
6. Sahrmann, S. *Diagnosis and Treatment of Movement Impairment Syndromes*, 1st ed.; Mosby: St. Louis, MO, USA, 2001; ISBN 978-0-8016-7205-7.
7. McCreary, E.K.; Provance, P.G.; Rodgers, M.M.; Romani, W.A.; Kendall, F.P. *Muscles: Testing and Function with Posture and Pain*, 5th ed.; Lippincott Williams & Wilkins: Baltimore, MD, USA, 2005; ISBN 978-0-7817-4780-6.
8. Janssen, M.M.A.; Kouwenhoven, J.-W.M.; Schlösser, T.P.C.; Viergever, M.A.; Bartels, L.W.; Castelein, R.M.; Vincken, K.L. Analysis of Preexistent Vertebral Rotation in the Normal Infantile, Juvenile, and Adolescent Spine. *Spine* **2011**, *36*, E486–E491. [CrossRef]
9. Dolphens, M.; Cagnie, B.; Coorevits, P.; Vleeming, A.; Palmans, T.; Danneels, L. Posture Class Prediction of Pre-Peak Height Velocity Subjects According to Gross Body Segment Orientations Using Linear Discriminant Analysis. *Eur. Spine J.* **2014**, *23*, 530–535. [CrossRef]
10. Tribus, C.B. Scheuermann's Kyphosis in Adolescents and Adults: Diagnosis and Management. *J. Am. Acad. Orthop. Surg.* **1998**, *6*, 36–43. [CrossRef]
11. Lowe, T.G. Scheuermann's Kyphosis. *Neurosurg. Clin. N. Am.* **2007**, *18*, 305–315. [CrossRef]
12. Mac-Thiong, J.-M.; Labelle, H.; Roussouly, P. Pediatric Sagittal Alignment. *Eur. Spine J.* **2011**, *20*, 586–590. [CrossRef]
13. Russo, L.; Giustino, V.; Toscano, R.E.; Secolo, G.; Secolo, I.; Iovane, A.; Messina, G. Can Tongue Position and Cervical ROM Affect Postural Oscillations? A Pilot and Preliminary Study. *J. Hum. Sport Exerc.* **2020**, *15*, S840–S847, In Proceedings of the Journal of Human Sport and Exercise-2020-Spring Conferences of Sports Science; Universidad de Alicante, 2020.
14. Russo, L.; Bartolucci, P.; Ardigò, L.P.; Padulo, J.; Pausic, J.; Iacono, A.D. An Exploratory Study on the Acute Effects of Proprioceptive Exercise and/or Neuromuscular Taping on Balance Performance. *Asian J. Sport. Med.* **2018**, *9*, e63020. [CrossRef]
15. Giustino, V.; Parroco, A.M.; Gennaro, A.; Musumeci, G.; Palma, A.; Battaglia, G. Physical Activity Levels and Related Energy Expenditure during COVID-19 Quarantine among the Sicilian Active Population: A Cross-Sectional Online Survey Study. *Sustainability* **2020**, *12*, 4356. [CrossRef]
16. Cohen, L.; Kobayashi, S.; Simic, M.; Dennis, S.; Refshauge, K.; Pappas, E. Non-Radiographic Methods of Measuring Global Sagittal Balance: A Systematic Review. *Scoliosis* **2017**, *12*, 30. [CrossRef]
17. Roggio, F.; Petrigna, L.; Trovato, B.; Zanghì, M.; Sortino, M.; Vitale, E.; Rapisarda, L.; Testa, G.; Pavone, V.; Pavone, P.; et al. Thermography and Rasterstereography as a Combined Infrared Method to Assess the Posture of Healthy Individuals. *Sci. Rep.* **2023**, *13*, 4263. [CrossRef]
18. Roggio, F.; Petrigna, L.; Filetti, V.; Vitale, E.; Rapisarda, V.; Musumeci, G. Infrared Thermography for the Evaluation of Adolescent and Juvenile Idiopathic Scoliosis: A Systematic Review. *J. Therm. Biol.* **2023**, *113*, 103524. [CrossRef] [PubMed]

19. Roggio, F.; Ravalli, S.; Maugeri, G.; Bianco, A.; Palma, A.; Di Rosa, M.; Musumeci, G. Technological Advancements in the Analysis of Human Motion and Posture Management through Digital Devices. *World J. Orthop.* **2021**, *12*, 467–484. [CrossRef]
20. Krott, N.L.; Wild, M.; Betsch, M. Meta-Analysis of the Validity and Reliability of Rasterstereographic Measurements of Spinal Posture. *Eur. Spine J.* **2020**, *29*, 2392–2401. [CrossRef]
21. Elsayed, W.; Farrag, A.; Muaidi, Q.; Almulhim, N. Relationship between Sagittal Spinal Curves Geometry and Isokinetic Trunk Muscle Strength in Adults. *Eur. Spine J.* **2018**, *27*, 2014–2022. [CrossRef]
22. Russo, L.; Padulo, J. Letter to the Editor Concerning "Relationship between Sagittal Spinal Curves Geometry and Isokinetic Trunk Muscle Strength in Adults" by Elsayed W, Farrag A, Muaidi Q, Almulhim N (Eur Spine J [2018] 27:2014–2022). *Eur Spine J.* **2019**, *28*, 191–192. [CrossRef]
23. Mohokum, M.; Mendoza, S.; Udo, W.; Sitter, H.; Paletta, J.R.; Skwara, A. Reproducibility of Rasterstereography for Kyphotic and Lordotic Angles, Trunk Length, and Trunk Inclination: A Reliability Study. *Spine* **2010**, *35*, 1353–1358. [CrossRef] [PubMed]
24. Molinaro, L.; Russo, L.; Cubelli, F.; Taborri, J.; Rossi, S. Reliability Analysis of an Innovative Technology for the Assessment of Spinal Abnormalities. In Proceedings of the 2022 IEEE International Symposium on Medical Measurements and Applications (MeMeA), Messina, Italy, 22–24 June 2022; pp. 1–6.
25. Roghani, T.; Khalkhali Zavieh, M.; Rahimi, A.; Talebian, S.; Dehghan Manshadi, F.; Akbarzadeh Baghban, A.; King, N.; Katzman, W. The Reliability of Standing Sagittal Measurements of Spinal Curvature and Range of Motion in Older Women With and Without Hyperkyphosis Using a Skin-Surface Device. *J. Manip. Physiol. Ther.* **2017**, *40*, 685–691. [CrossRef] [PubMed]
26. Barrett, E.; McCreesh, K.; Lewis, J. Reliability and Validity of Non-Radiographic Methods of Thoracic Kyphosis Measurement: A Systematic Review. *Man. Ther.* **2014**, *19*, 10–17. [CrossRef] [PubMed]
27. Kellis, E.; Adamou, G.; Tzilios, G.; Emmanouilidou, M. Reliability of Spinal Range of Motion in Healthy Boys Using a Skin-Surface Device. *J. Manip. Physiol. Ther.* **2008**, *31*, 570–576. [CrossRef] [PubMed]
28. Mannion, A.F.; Knecht, K.; Balaban, G.; Dvorak, J.; Grob, D. A New Skin-Surface Device for Measuring the Curvature and Global and Segmental Ranges of Motion of the Spine: Reliability of Measurements and Comparison with Data Reviewed from the Literature. *Eur. Spine J.* **2004**, *13*, 122–136. [CrossRef]
29. Belli, G.; Toselli, S.; Latessa, P.M.; Mauro, M. Evaluation of Self-Perceived Body Image in Adolescents with Mild Idiopathic Scoliosis. *Eur. J. Investig. Health Psychol. Educ.* **2022**, *12*, 319–333. [CrossRef]
30. Macedo Ribeiro, A.F.; Bergmann, A.; Lemos, T.; Pacheco, A.G.; Mello Russo, M.; Santos de Oliveira, L.A.; de Carvalho Rodrigues, E. Reference Values for Human Posture Measurements Based on Computerized Photogrammetry: A Systematic Review. *J. Manip. Physiol. Ther.* **2017**, *40*, 156–168. [CrossRef]
31. Krawczky, B.; Pacheco, A.G.; Mainenti, M.R.M. A Systematic Review of the Angular Values Obtained by Computerized Photogrammetry in Sagittal Plane: A Proposal for Reference Values. *J. Manip. Physiol. Ther.* **2014**, *37*, 269–275. [CrossRef]
32. Trovato, B.; Roggio, F.; Sortino, M.; Zanghì, M.; Petrigna, L.; Giuffrida, R.; Musumeci, G. Postural Evaluation in Young Healthy Adults through a Digital and Reproducible Method. *J. Funct. Morphol. Kinesiol.* **2022**, *7*, 98. [CrossRef]
33. Russo, L.; Montagnani, E.; Pietrantuono, D.; D'Angona, F.; Fratini, T.; Di Giminiani, R.; Palermi, S.; Ceccarini, F.; Migliaccio, G.M.; Lupu, E.; et al. Self-Myofascial Release of the Foot Plantar Surface: The Effects of a Single Exercise Session on the Posterior Muscular Chain Flexibility after One Hour. *Int. J. Environ. Res. Public Health* **2023**, *20*, 974. [CrossRef]
34. Porto, A.B.; Okazaki, V.H.A. Procedures of Assessment on the Quantification of Thoracic Kyphosis and Lumbar Lordosis by Radiography and Photogrammetry: A Literature Review. *J. Bodyw. Mov. Ther.* **2017**, *21*, 986–994. [CrossRef] [PubMed]
35. Porto, A.B.; Okazaki, V.H.A. Thoracic Kyphosis and Lumbar Lordosis Assessment by Radiography and Photogrammetry: A Review of Normative Values and Reliability. *J. Manip. Physiol. Ther.* **2018**, *41*, 712–723. [CrossRef] [PubMed]
36. Post, R.B.; Leferink, V.J.M. Spinal Mobility: Sagittal Range of Motion Measured with the SpinalMouse, a New Non-Invasive Device. *Arch. Orthop. Trauma Surg.* **2004**, *124*, 187–192. [CrossRef]
37. Livanelioglu, A.; Kaya, F.; Nabiyev, V.; Demirkiran, G.; Fırat, T. The Validity and Reliability of "Spinal Mouse" Assessment of Spinal Curvatures in the Frontal Plane in Pediatric Adolescent Idiopathic Thoraco-Lumbar Curves. *Eur. Spine J.* **2016**, *25*, 476–482. [CrossRef] [PubMed]
38. Ardigò, L.P.; Palermi, S.; Padulo, J.; Dhahbi, W.; Russo, L.; Linetti, S.; Cular, D.; Tomljanovic, M. External Responsiveness of the SuperOpTM Device to Assess Recovery After Exercise: A Pilot Study. *Front. Sport. Act. Living* **2020**, *2*, 67. [CrossRef] [PubMed]
39. Russo, L.; Di Capua, R.; Arnone, B.; Borrelli, M.; Coppola, R.; Esposito, F.; Padulo, J. Shoes and Insoles: The Influence on Motor Tasks Related to Walking Gait Variability and Stability. *Int. J. Environ. Res. Public Health* **2020**, *17*, 4569. [CrossRef]
40. Singla, D.; Veqar, Z.; Hussain, M.E. Photogrammetric Assessment of Upper Body Posture Using Postural Angles: A Literature Review. *J. Chiropr. Med.* **2017**, *16*, 131–138. [CrossRef]
41. Saad, K.R.; Colombo, A.S.; Ribeiro, A.P.; João, S.M.A. Reliability of Photogrammetry in the Evaluation of the Postural Aspects of Individuals with Structural Scoliosis. *J. Bodyw. Mov. Ther.* **2012**, *16*, 210–216. [CrossRef]
42. Azadinia, F.; Hosseinabadi, M.; Ebrahimi, I.; Mohseni-Bandpei, M.-A.; Ghandhari, H.; Yassin, M.; Behtash, H.; Ganjavian, M.-S. Validity and Test–Retest Reliability of Photogrammetry in Adolescents with Hyperkyphosis. *Physiother. Theory Pract.* **2022**, *38*, 3018–3026. [CrossRef]
43. Stolinski, L.; Kozinoga, M.; Czaprowski, D.; Tyrakowski, M.; Cerny, P.; Suzuki, N.; Kotwicki, T. Two-Dimensional Digital Photography for Child Body Posture Evaluation: Standardized Technique, Reliable Parameters and Normative Data for Age 7–10 Years. *Scoliosis* **2017**, *12*, 38. [CrossRef]

44. Dodge, Y. *The Concise Encyclopedia of Statistics*, 2010th ed.; Springer: Philadelphia, PA, USA, 2008; ISBN 978-0-397-51837-1.

45. Bland, J.M.; Altman, D.G. Statistical Methods for Assessing Agreement between Two Methods of Clinical Measurement. *Lancet* **1986**, *1*, 307–310. [CrossRef] [PubMed]

46. Cirillo Totera, J.I.; Fleiderman Valenzuela, J.G.; Garrido Arancibia, J.A.; Pantoja Contreras, S.T.; Beaulieu Lalanne, L.; Alvarez-Lemos, F.L. Sagittal Balance: From Theory to Clinical Practice. *EFORT Open Rev.* **2021**, *6*, 1193–1202. [CrossRef] [PubMed]

47. Demir, E.; Güzel, N.; Çobanoğlu, G.; Kafa, N. The Reliability of Measurements with the Spinal Mouse Device in Frontal and Sagittal Planes in Asymptomatic Female Adolescents. *Ann. Clin. Anal. Med.* **2020**, *11*, 146–149. [CrossRef]

48. Muyor, J.M.; López-Miñarro, P.A.; Alacid, F. The Relationship between Hamstring Muscle Extensibility and Spinal Postures Varies with the Degree of Knee Extension. *J. Appl. Biomech.* **2013**, *29*, 678–686. [CrossRef]

49. Furlanetto, T.S.; Sedrez, J.A.; Candotti, C.T.; Loss, J.F. Photogrammetry as a Tool for the Postural Evaluation of the Spine: A Systematic Review. *World J. Orthop.* **2016**, *7*, 136–148. [CrossRef]

50. López-Miñarro, P.A.; Vaquero-Cristóbal, R.; Alacid, F.; Isorna, M.; Muyor, J.M. Comparison of Sagittal Spinal Curvatures and Pelvic Tilt in Highly Trained Athletes from Different Sport Disciplines. *Kinesiology* **2017**, *49*, 109–116. [CrossRef]

51. Rabieezadeh, A.; Hovanloo, F.; Khaleghi, M.; Akbari, H. The relationship of height, weight and body mass index with curvature of spine kyphosis and lordosis in 12–15-year old male adolescents of Tehran. *Turk. J. Sport Exe.* **2016**, *18*, 42–46.

52. Hazar, Z.; Karabicak, G.O.; Tiftikci, U. Reliability of Photographic Posture Analysis of Adolescents. *J. Phys. Ther. Sci.* **2015**, *27*, 3123–3126. [CrossRef]

53. Carregaro, R.L.; Silva, L.; Gil Coury, H.J.C. Comparison between Two Clinical Tests for Evaluating the Flexibility of the Posterior Muscles of the Thigh. *Braz. J. Phys. Ther.* **2007**, *11*, 139–145. [CrossRef]

54. Navarro, I.J.R.L.; Candotti, C.T.; do Amaral, M.A.; Dutra, V.H.; Gelain, G.M.; Loss, J.F. Validation of the Measurement of the Angle of Trunk Rotation in Photogrammetry. *J. Manip. Physiol. Ther.* **2020**, *43*, 50–56. [CrossRef]

Journal of
Functional Morphology and Kinesiology

Article

Effects of Feedback-Supported Online Training during the Coronavirus Lockdown on Posture in Children and Adolescents

Oliver Ludwig [1,*], Carlo Dindorf [1], Torsten Schuh [2], Thomas Haab [3], Johannes Marchetti [1,3] and Michael Fröhlich [1]

[1] Department of Sports Science, Technische Universität Kaiserslautern, 67663 Kaiserslautern, Germany
[2] Sport Performance Education, 66386 Sankt Ingbert, Germany
[3] Institute of Sports Science, Saarland University, 66123 Saarbrücken, Germany
* Correspondence: oliver.ludwig@sowi.uni-kl.de

Abstract: (1) Background. The coronavirus pandemic had a serious impact on the everyday life of children and young people with sometimes drastic effects on daily physical activity time that could have led to posture imbalances. The aim of the study was to examine whether a six-week, feedback-supported online training programme could improve posture parameters in young soccer players. (2) Methods. Data of 170 adolescent soccer players (age 15.6 ± 1.6 years) were analyzed. A total of 86 soccer players of a youth academy participated in an online training program that included eight exercises twice per week for 45 min (Zoom group). The participants' exercise execution could be monitored and corrected via smartphone or laptop camera. Before and after the training intervention, participants' posture was assessed using photographic analysis. The changes of relevant posture parameters (perpendicular positions of ear, shoulder and hips, pelvic tilt, trunk tilt and sacral angle) were statistically tested by robust mixed ANOVA using trimmed means. Postural parameters were also assessed post hoc at 8-week intervals in a control group of 84 participants of the same age. (3) Results. We found a statistically significant interaction ($p < 0.05$) between time and group for trunk tilt, head and shoulder protrusion and for hip anteversion in the Zoom group. No changes were found for these parameters in the control group. For pelvic tilt no significant changes were found. (4) Conclusions. Feedback-based online training with two 45 min sessions per week can improve postural parameters in adolescent soccer players over a period of six weeks.

Keywords: posture training; feedback; COVID-19; neck-shoulder-region; shoulder protraction; upper crossed syndrome; posture weakness; physical inactivity; sedentary behavior; soccer

Citation: Ludwig, O.; Dindorf, C.; Schuh, T.; Haab, T.; Marchetti, J.; Fröhlich, M. Effects of Feedback-Supported Online Training during the Coronavirus Lockdown on Posture in Children and Adolescents. *J. Funct. Morphol. Kinesiol.* **2022**, *7*, 88. https://doi.org/10.3390/jfmk7040088

Academic Editor: Peter Hofmann

Received: 26 September 2022
Accepted: 13 October 2022
Published: 14 October 2022

Publisher's Note: MDPI stays neutral with regard to jurisdictional claims in published maps and institutional affiliations.

1. Introduction

Postural disorders are widespread in childhood and adolescence and are reported with a prevalence of 25–60% [1–3]. A protracted shoulder girdle, a protracted head, a hollow back, and a hunchback are the most common postural weaknesses in adolescents [4]. Current studies have shown that this can lead to pain at an older age. For example, Dolphens et al. demonstrated in 1196 children and adolescents that increased forward head tilt was associated with lifetime prevalence of neck pain [5]. Kim et al. in turn were able to demonstrate, that a protracted head could lower respiratory functions in young adults [6]. An important cause for the occurrence of postural weaknesses is apparently muscular imbalances that lead to the displacement of individual body parts [7,8]. A sedentary everyday life seems to promote such imbalances, and the intensive use of smartphones also apparently leads to postural adaptations, especially forward displacement of the shoulder and head [9,10].

In 2020, the coronavirus pandemic had a serious impact on the everyday life of children and young people through measures such as lockdowns and temporary curfews. In Germany, as in many other countries around the world, young people who were normally active in sports were cut off from outdoor exercise and sports activities in clubs for several months. While physical activity time thus decreased significantly, time in front of the computer increased—as a leisure activity, but also because school lessons were largely held online. This promoted physical inactivity among many adolescents in Germany and worldwide for weeks, with sometimes drastic effects on daily physical activity time. For example, Schmidt et al. showed that in Germany, among 14–17-year-old boys, physical activity decreased by an average of 21 min per day during Corona, while time spent in front of the screen relaxing increased by 79 min per day [11]. A survey of over 6400 children and adolescents in France by Chambonniere et al. also showed that 57% of teenagers significantly decreased their physical activity during Corona, which was particularly evident in adolescents who were previously more active in sports [12]. Guan and colleagues found the same trend in Canadian adolescents, namely, in addition to a decrease in physical activity time, a 66% increase in time spent watching television and a 35% increase in time spent playing video games [13].

It is well known that there is a strong association between health outcomes and physical activity behavior. Guan et al. therefore concluded that the health of children and adolescents may have been compromised due to a lack of physical activity during the Corona pandemic [13].

Negative effects of sitting on altered muscular activity and misalignments of the head and neck have been demonstrated. For example, Caneiro et al. found the typical postural deficiencies in the form of increased neck flexion and increased anterior translation of the head as a result of a slump sitting position, which led to an unfavorable increase in the activity of the postural muscles in the neck region [14], while Ertekin and Günaydin demonstrated a change in muscle activity and stiffness of the trapezius muscle associated with protracted shoulders [15].

Many of these muscularly induced misalignments are associated with neck pain. For example, Ruivo et al. observed protracted head in 68% of the subjects and protracted shoulders in 58% of the subjects in studies of 275 adolescents, and forward head posture was significantly associated with neck pain [16].

According to current scientific knowledge, an important starting point in the prevention of shoulder and neck pain is the strengthening of the stabilizing musculature [17]. Targeted sports activities can contribute to this.

One approach during the pandemic time, which was physically and psychologically stressful for children and adolescents, was sports activities carried out online in groups. This is where the present study comes in. For adolescent soccer players at a DFB (German Football Association) youth academy, who normally trained together three to four times a week on the sports field, posture-specific sports training was offered online twice a week via the Zoom® platform during the lockdown phase. The typical postural weaknesses of this target group were known from previous own studies [8], so the training programme was adapted accordingly. As the online meetings took place live via smartphone or laptop cameras, the exercises could be individually corrected by trainers in real time for each participant.

Although some studies discuss the general use of digital tools and online training or effectiveness regarding improvements in overall activity levels during the coronavirus pandemic, there is insufficient research on the effectiveness of online training on body posture. Tjønndal was able to show that the use of online tools during the corona pandemic was quite uncoordinated and that synchronized (live-streamed) training was offered rather rarely [18]. On the other hand, when Parker et al. surveyed 963 adolescents, they found that the willingness to use digital platforms to support and guide physical activity was present in 26.5% of the surveyed adolescents [19]. These are initially good conditions, especially since 95–96% of 13–18-year-olds in Germany had their own smartphone at the end of 2021 [20].

However, there is very little research on the use and effectiveness of synchronized (live streamed) online training, particularly for posture correction and especially when participants received live feedback. Given the potential and benefits in situations where face-to-face training cannot take place, there is a clear need for research. Therefore, the aim of the study was to examine whether a six-week, feedback-supported online training programme could improve posture parameters in children and adolescents.

2. Materials and Methods

The online training intervention took place during the second "hard" coronavirus lockdown in Germany, which came into effect on 16 December 2020 and lasted for six weeks.

2.1. Subjects

A total of 96 adolescents from the U14 (age 13–14 years) to U19 (age 17–19 years) youth teams of the DFB youth academy of the SV 07 Elversberg soccer club (Germany) participated in the training intervention (Zoom group). Participation was voluntary and was advertised via the club. The data of 86 participants of the Zoom group were used for final analysis (anthropometric data in Table 1); data of ten subjects had to be removed because regular participation was not possible (due to poor internet connection or illness) or the follow-up appointment was missed.

Table 1. Anthropometric data of the subjects.

	Zoom (N = 86)	Control (N = 84)
Age [years]	15.6 ± 1.6	15.7 ± 1.6
Weight [kg]	63.0 ± 12.2	65.3 ± 10.9
Height [cm]	173.0 ± 9.8	174.1 ± 9.0

Since, for ethical reasons, we could not deprive young soccer players of the club of participating in online training during the coronavirus lockdown, no non-training control group could run in parallel. We therefore selected an age-matched control group of 84 adolescent soccer players post hoc, who were examined twice at intervals of eight weeks and only took part in regular soccer training or had a training break in between.

All participants or their legal representatives were fully informed beforehand and gave their written informed consent to participate in the postural analyses and to take part in the video conferences. The study was conducted in accordance with the Declaration of Helsinki and approved by the local University Ethics Committee (ref. nr. 6-18).

2.2. Posture Analysis

Before the start of the study, a posture analysis was carried out on all participants on the premises of the association. Due to the contact restrictions, an exemption was obtained from the responsible municipal regulatory authority. In view of the current infection situation, appropriate corona precaution measures were followed strictly.

During the analysis, the subjects were first weighed in their underwear, and their height was determined (SECA stadiometer). The following anatomical landmarks were marked with marker dots or marker balls (diameter 12 mm): malleolus lateralis, greater trochanter, acromion, spinous process of the 7th cervical vertebra (C7), spinous process of the 1st sacral vertebra (S1), anterior superior iliac spine (ASIS), and posterior superior iliac spine (PSIS). The subjects then placed themselves in front of a measuring wall and assumed a relaxed posture (arms hanging loosely, looking straight ahead, breathing normally). A posture photo was taken using a camera (Logitech webcam, Full HD) mounted on a tripod at hip height. The inclination of the sacrum was then determined in degrees by placing an electronic inclinometer (Neoteck NTK033) on the proximal part of the sacrum.

Using analysis software (Dartfish ProSuite 6, Dartfish, Fribourg, Switzerland), the following posture parameters were determined (Figure 1): horizontal distances of the ear, shoulder (acromion), and hip (major trochanter) perpendicularly through the lateral malleolus as a percentage of body height; pelvic tilt (angle between the connecting the

PSIS–ASIS line to the horizontal); and trunk forward tilt (angle between the connecting the S1–C7 line to the vertical).

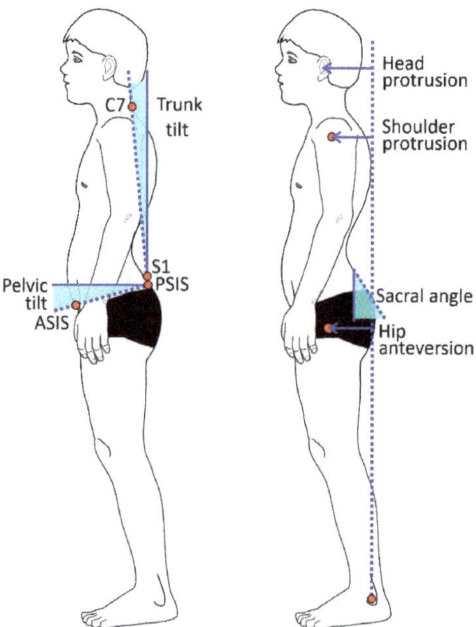

Figure 1. Analyzed posture parameters; ASIS = anterior superior iliac spine, PSIS = posterior superior iliac spine, C7 = 7th cervical vertebrae, S1 = 1st sacral vertebrae.

Posture assessment using analysis of angles and distances from posture photographs is a scientifically well-studied, reliable and valid procedure [21,22].

2.3. Intervention

To implement the intervention, a temporary film studio was set up in the club's athletics hall, from which the training exercises were broadcast live twice a week via the Zoom® (Zoom Video Communications Inc., San José, CA, USA) communications platform. One trainer performed the exercises at the designated pace, while a second trainer commented on the execution of the exercises and announced the number of repetitions and rest times. The participants had previously been asked to point their smartphone or laptop camera so that they were clearly visible in the front or side profile. Two other trainers switched from participant to participant via video during the exercise session and controlled and, where necessary, verbally corrected the execution of the exercise until the participant executed it correctly (Figure 2). The participants were divided into a total of three subgroups of 30–35 adolescents, who took part in the online training at different times to ensure sufficient monitoring by the trainers. The first two sessions were necessary to learn how to perform the exercises correctly.

From the postural examinations (pre-test), it was known that there were deficits especially in the head and shoulder area (protracted shoulder girdle, head protrusion), as well as in the pelvic area (Table 2). Therefore, the exercises were chosen to meet the following criteria:

1. Strengthening of the target muscles for segmental postural improvement;
2. Feasibility with aids available in the household;
3. Easy to correct via video.

Figure 2. Exemplary online view that the participants (**a**) and the correcting trainers (**b**) had during the exercise sessions. In the photo on the right, the background has been changed for data protection reasons. The persons depicted have given their written consent to the publication of the photos.

Table 2. Prevalence of postural deviations in the studied population (absolute N/prevalence).

Age	N	Forward Head	Protracted Shoulder	Scapulae Alatae	Pelvic Anteversion	Increased Pelvic Tilt	Hunchback
13–14	17	8/47%	10/59%	3/18%	2/12%	8/47%	2/12%
14–15	17	11/65%	12/71%	1/6%	11/65%	3/18%	7/41%
15–16	20	11/55%	11/55%	3/15%	11/55%	4/20%	2/10%
16–17	20	8/40%	7/35%	2/10%	10/50%	1/5%	5/25%
17–19	22	14/64%	11/50%	1/5%	15/68%	3/14%	5/23%
Sum	96	52/54%	51/53%	10/10%	49/51%	19/20%	21/22%

To improve the position of the shoulder and head, particular emphasis was placed on strengthening the rhomboid, latissimus dorsi, rectus capitis, longus colli, longus capitis, and obliquus capitis muscles [23]. To correct the pelvic position in the sense of reducing the pelvic tilt, the rectus abdominis, gluteus maximus, and ischiocrural muscle groups were strengthened [24]. The exercises performed are described in Table 3.

Table 3. Description of the exercises and the targeted muscle groups.

Exercise	Target Muscles	Execution
#1 Swimmer	- M. deltoideus pars spinalis - M. rhomboideus major and minor - M. trapezius pars ascendens - M. erector spinae - M. gluteus maximus - M. biceps femoris - M. longus colli - M. longus capitis	*Starting position:* Prone, arms and legs extended, feet slightly off the floor, head slightly raised, looking down towards the floor. *Execution:* Bring arms back in a motion similar to breaststroke with the thumb pointing upwards. Raise the arms above bottles placed at the sides at shoulder height and then lower them again. Consciously pull the shoulder blades together. Keep the legs in a straight position.
#2 Reverse Butterfly in Standing Position	- M. deltoideus pars spinalis - M. trapezius pars ascendens - M. rhomboideus major and minor	*Starting position:* Stand hip-width apart, knees slightly bent, upper body bent forward. Keep back straight throughout the exercise. Hold dumbbells/bottles in upper grip and bend arms slightly. *Execution:* Raise arms sideways until elbows are level with shoulders, contracting scapulae. Slowly lower the arms to the sides.

Table 3. *Cont.*

Exercise	Target Muscles	Execution
#3 Torso Rotation with Lunge Backward	- M. obliquus internus abdominis - M. obliquus externus abdominis - M. transversus abdominis - M. deltoideus pars clavicularis	*Starting position:* Lunge backwards, hold exercise ball in front of the body with arms outstretched, tense abdomen. *Execution:* Slowly rotate the upper body as far as possible, always looking at the ball, hold briefly and then rotate to the other side.
#4 Push-Ups on the Ball	- M. rectus abdominis - M. transversus abdominis - M. obliquus internus abdominis - M. obliquus externus abdominis	*Starting position:* Lean on the exercise ball with bent forearms; upper body, buttocks and legs in a straight line, keep tension. *Execution:* Lower arms perform a circular movement with the ball, keep body tension.
#5 Rowing with Exercise Band in Standing Position	- M. latissimus dorsi - M. rhomboideus minor et major - M. trapezius pars ascendens, descendes und transversa	*Starting position:* Stand hip-width apart, knees slightly bent, upper body bent forward. Back is straight. Hold the exercise band around the feet with both hands, arms slightly bent. *Execution:* Draw shoulder blades together, then tighten the exercise band with bent arms, hold position briefly, then slowly return to starting position.
#6 Pelvic lift with Exercise Ball	- M. gluteus maximus - M. erector spinae pars lumbalis - M. deltoideus	*Starting position:* Supine, heels on the exercise ball, body forms a straight line and has tension, only shoulders and arms support the body, head lies relaxed on the floor in extension of the spine. *Execution:* Heels are pulled towards the buttocks, pelvis is lifted up to 90° flexion in the knee joint, then slowly return to starting position.
#7 Squat with Bar	- M. gluteus maximus - M. quadriceps femoris - M. erector spinae - M. deltoideus	*Starting position:* Feet slightly wider than shoulders, knees slightly out, hold broomstick above head with arms extended. *Execution:* Lower buttocks backwards in a controlled manner, back remains straight, do not push knees over toes, always turn knees slightly outwards, bend until 90° flexion in knee joint, then return to starting position.
#8 Neck Press	- M. longus colli - M. longus capitis - M. obliquus capitis	*Starting position:* Sitting on the floor, training band around the back of the head, holding both ends with the hands, upper body upright, back straight, hands at forehead level in front of the head. *Execution:* Head is moved backwards slowly and in a controlled manner.

The exercise sessions were performed twice a week in the afternoon, three days apart, and lasted about 45 min each. At the beginning, the participants completed an approximately five-minute warm-up programme consisting of running on the spot, jumping jacks, knee bends, and light stretching exercises for the trunk and leg muscles. This was followed by the eight exercises with 20 repetitions each (Table 3). This sequence was repeated a total of three times with short drinking breaks. The training was concluded with five minutes of stretching exercises for the trunk and legs and a check of the load intensity.

The exercises were demonstrated by a trainer in real time and, thus, the number of repetitions, load, and rest times were specified. An adjustment to the individual performance level of the participants was made through the choice of weights (e.g., water bottles of different sizes), the height of the obstacles (for the "swimmer" exercise), and the tension of the resistance band (e.g., for rowing exercises). In the first week of training, the correct execution of the movement was trained and the weights were adjusted so that the number of repetitions could be carried out. After each training session, the participants were asked to rate the training load on a ten-point Likert scale via chat. The target range was set at 6–8; if the target was exceeded or not reached, the resistance and weights were varied individually in consultation with the participants. After seven training sessions, either the weights or the number of repetitions (now 30) was increased. This was maintained until the 12th training session.

2.4. Statistics

Outliers at the time of the pre-tests were checked again in the posture photos. As these were not measurement errors but actual extreme values in the range of the posture parameters, they were not removed.

Due to presence of outliers, as well as partial inhomogeneity of the covariance according to Box's test ($p < 0.01$), assumptions could not be met for the calculation of a commonly used mixed ANOVA. Therefore, a robust mixed ANOVA using trimmed means was applied for statistical analysis of the interaction effects of the groups and measuring points in time using the R package WRS2 [25]. Further post hoc statistical analysis was performed using IBS SPSS Statistics (version 16, SPSS Inc., Chicago, IL, USA). Visualizations were performed using the Python library "Seaborn" [26].

3. Results

Table 4 shows the development of the posture parameters in the training and control groups, while Figure 3 shows the distribution and development of the parameters over time.

3.1. Interaction Effects

There was a statistically significant interaction between time and group for head protrusion ($F(1, 93.75) = 51.92$, $p < 0.001$, partial $\eta^2 = 0.36$), for shoulder protrusion ($F(1, 93.85) = 23.62$, $p < 0.001$, partial $\eta^2 = 0.20$), and for hip anteversion ($F(1, 100.99) = 7.07$, $p = 0.009$, partial $\eta^2 = 0.07$).

We found a statistically significant interaction between time and group for trunk tilt ($F(1, 96.03) = 40.65$, $p < 0.001$, partial $\eta^2 = 0.30$).

For pelvic tilt, we found no statistically significant interaction between time and group ($F(1, 99.43) = 1.93$, $p = 0.168$). Further, no main effects for time ($F(1, 92.49) = 0.15$, $p = 0.695$) or group ($F(1, 99.43) = 0.48$, $p = 0.492$) could be found.

There was no statistically significant interaction between time and group for sacral angle ($F(1, 100.99) = 2.44$, $p < 0.001$). There were, however, main effects for time ($F(1, 97.96) = 28.64$, $p < 0.001$, partial $\eta^2 = 0.23$) and group ($F(1, 100.99) = 48.94$, $p < 0.001$, partial $\eta^2 = 0.33$).

3.2. Main Effects of "Between-Subject Factors" (Group)

The control and training groups did not differ significantly at the pre-test on the variables head protrusion, shoulder protrusion, hip anteversion, and pelvic tilt ($p > 0.05$), but there were differences in trunk tilt and sacral angle ($p < 0.05$). At the post-test, the training and control groups differed significantly ($p < 0.05$) in all the variables except pelvic tilt ($p > 0.05$).

3.3. Main Effects of "Within-Subject Factors" (Time)

For the training group, there was a significant change between the pre- and post-tests for all parameters ($p < 0.05$), except for pelvic tilt. For the control group, there was no difference between the measurement times ($p > 0.05$), except for the variables sacral angle and trunk tilt ($p < 0.05$).

Table 4. Development of the posture parameters over time (mean ± standard deviation).

Group Time	Zoom		Control	
	Pre	Post	Pre	Post
Head protrusion [% BH]	4.4 ± 1.8	1.8 ± 1.3	4.0 ± 1.6	3.9 ± 1.5
Shoulder protrusion [% BH]	3.1 ± 1.8	1.4 ± 1.0	3.2 ± 1.5	3.1 ± 1.6
Hip anteversion [% BH]	3.1 ± 1.4	2.2 ± 1.2	3.1 ± 1.1	3.3 ± 4.7
Trunk tilt [°]	3.6 ± 1.8	2.1 ± 1.3	2.8 ± 1.3	3.6 ± 1.4
Pelvic tilt [°]	9.5 ± 5.6	9.7 ± 3.6	9.9 ± 3.7	9.7 ± 4.4
Sacrum angle [°]	21.8 ± 7.9	17.9 ± 6.0	17.8 ± 5.8	12.4 ± 5.9

% BH = percentage of body height.

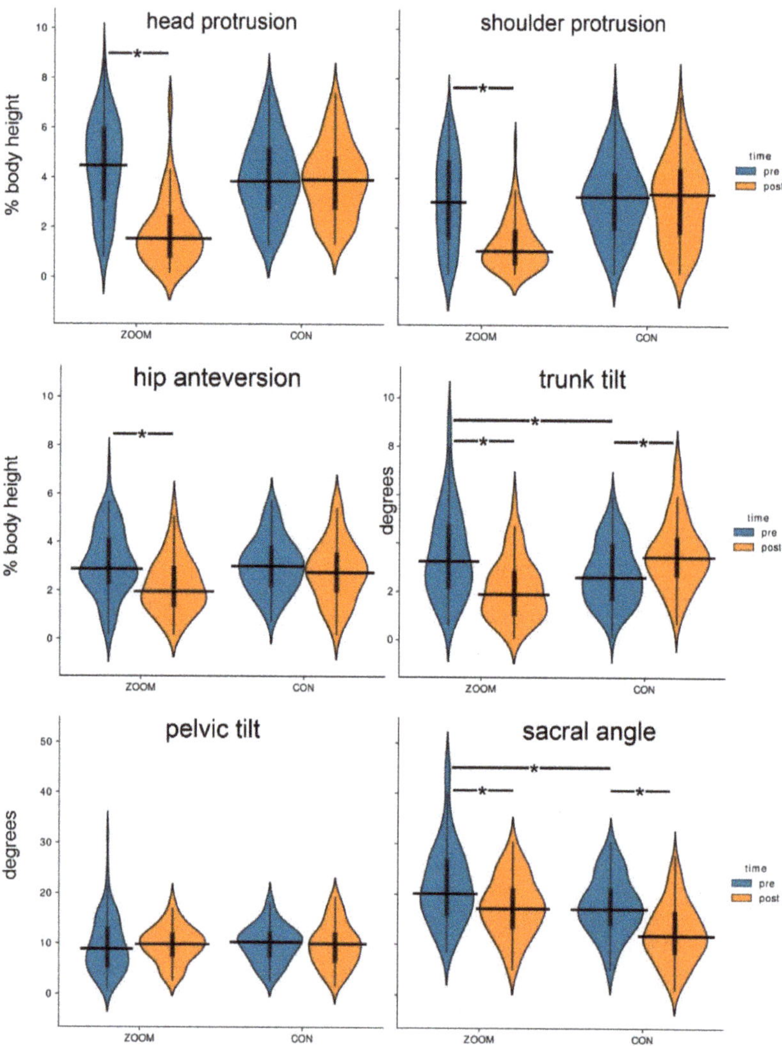

Figure 3. Violin plots of the development of the posture parameters of the training group (ZOOM) and the control group (CON). The horizontal lines mark the median, * = significant difference ($p < 0.05$).

4. Discussion

The coronavirus pandemic brought drastic changes to the everyday lives of many people, especially during the phases of lockdown, with contact restrictions and massive reductions in social and sporting life. Children and adolescents were particularly affected [11]. The subjects of our study were young soccer players in the junior performance sector who previously trained three to four times a week for 90 min or more with an additional game at the weekend. During the phases of the second hard lockdown, which lasted from mid-December 2020 to May 2021 in Germany, club life was largely switched off. Training sessions conducted at home and accompanied online were therefore the only guided sporting activity for many young people and, at the same time, one of the few opportunities to maintain social contacts within the team.

The present study was able to show that significant improvements in posture could be brought about by the six-week online training programme. To substantiate the study results, the changes in posture parameters during the six weeks of training were compared post hoc with the posture development of a roughly identically composed group of young competitive soccer players in regular sporting activity after the coronavirus lockdown phase. Although a control group running at the same time would have been scientifically optimal, it could not be realised at the time the study was conducted for ethical reasons, as we did not want to deny any young person participation in the exercise programme during the lockdown. The absence of systematic differences between both groups, and, therefore, the potential suitability, is also indicated by the mostly non-significant differences for the pre-test condition.

The reduction in the perpendicular distances of the ear, the shoulder, and the hip show that an improvement in posture in the sense of a less forward-leaning position could be achieved, presumably through the training-related strengthening of the dorsal muscle chain. Figure 4 shows a typical example of the change in a participant's habitual posture over the course of the study. It is known that deviations of the head and trunk from the perpendicular can be associated with pain in the neck and back [27]. The violin plots in Figure 4 provide information about the distribution of these parameters and show that there was no normal distribution in the pre-tests, but rather a skewed distribution in the direction of the perpendicular. Minimum perpendicular distances are considered biomechanically optimal.

As a measure against protracted head, the longus colli, obliquus capitis, and longus capitis muscles were strengthened via Exercises 1 and 8 (Table 3). For retraction of the shoulders, it was additionally and synergistically important to strengthen the rhomboid, latissimus dorsi, and trapezius muscles, which were activated in Exercises 1, 2, and 5. This is consistent with the findings of Ruivo et al. [28], who were able to improve head and shoulder position in adolescents through a 16-week strength and stretching programme, and also with the work of Sheikhhoseini et al. [29], who were able to demonstrate the effect of appropriate interventions on head position.

Even though the training group and the control group differed in the parameter trunk tilt at the time of the pre-test, an improvement was observed in the training group in the post-test, but with even a deterioration in the control group. To reduce trunk tilt, the dorsal muscle chain (especially the gluteus maximus, erector spinae, and biceps femoris muscles) had to be strengthened (Exercises 5, 6, 7 in Table 3). The work of Dolphens et al. [5,30] showed that a forward-leaning posture in adolescence is an important predictor of back pain in the following years.

In the area of pelvic alignment, we could not find any significant improvements in the intervention group. A lifting of the anterior pelvic edge (reduction in the pelvic tilt and decrease in the sacral angle) would be associated with a reduction in lumbar lordosis after a new muscular balance between the inserting muscle groups has been established [7]. Strengthening the muscles that straighten the pelvis (especially the rectus abdominis, gluteus maximus, and hamstrings) would have been responsible for this. The training programme specifically included exercises for this purpose (1, 3, 4, 6, 7; Table 3). The fact that the training intervention did not bring about any improvement in this area could be related to the fact that a change in the pelvic position in the resting posture can presumably be brought about less through an increase in the strength of the muscles involved and more through an improvement in the proprioceptive perception of the pelvic position [24,31]. However, proprioceptive exercises were not part of the training programme. Furthermore, it cannot be ruled out that the training stimuli on the muscle groups mentioned were too low.

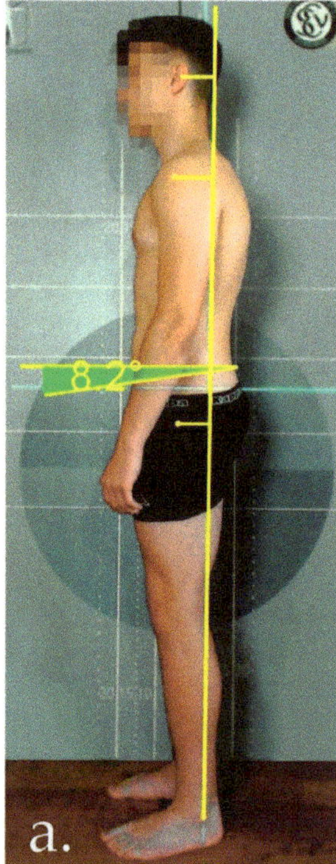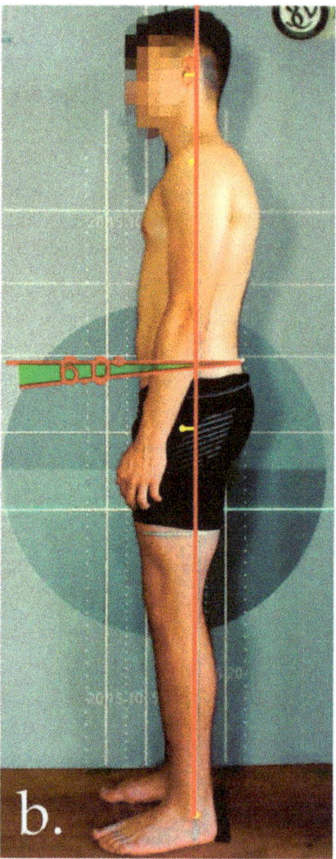

Figure 4. Typical example of the change in a participant's habitual posture (**a**: pre-test; **b**: post-test).

The control group did not show any improvement in the posture parameters (except for the sacral angle, which also decreased in the training group), so we assume that the online training was causally responsible for the improvement in the training group. This seems relevant to us insofar as the positive effects of targeted strength training on posture, which have already been proven in other studies [24,32–34], could also be achieved via online feedback training.

We see a significant aspect that contributed to the success and motivation of the participants in the feedback we were able to give in real time via video. It has been shown that in this way, frequent mistakes could be corrected, which otherwise would probably have led to a lower training success. At the same time, this kind of contact was an important psychological factor for many young people, as they reported afterwards. Praise and motivational announcements helped many participants to persevere through the strenuous training sessions and to maintain their bond with the team during the coronavirus period. From a scientific point of view, the lockdown situation naturally also had the advantage that (apart from individual sport at home) there were few outside influences in sporting terms, i.e., there was almost a "laboratory situation".

We see online feedback as a great opportunity for internet training, which is clearly superior to traditional training videos that are available in large numbers via social media and platforms such as YouTube® but offer no feedback option [18].

Our study has several limitations. As already mentioned, the intervention and control groups could not run in parallel. Nevertheless, the control group was composed as homogeneously as possible to the training group, even though they underwent regular sports training. However, since posture-relevant exercises were not part of their soccer training, we can exclude strong influencing effects here. It must also be emphasised that, apart from the load control, the training programme could not be individually adapted. Participants without postural deficits also performed the same exercises, so that a smaller effect of the intervention could be assumed here, which, in the end, rather led to an underestimation of the training effect. Nevertheless, training the postural muscles is important and makes sense in these cases as well, since many of the trained muscle groups also play a major role in movement stabilisation in soccer. In the selection of exercises, the choice of tools was limited to those available to the participants at home (water bottles as a substitute for dumbbells, exercise balls, resistance bands, broomsticks). Even though the adolescents had the specification not to do any additional training at home, this was not a factor that could be realistically controlled, but would not diminish the basic statement.

Our study was limited to performance-oriented youth soccer players. Nevertheless, we assume that the approach of feedback-supported training could also be successful for other youth target groups.

5. Conclusions

In the context of our study, it has been shown that it is possible to provide individual exercise supervision and correction for participants during online posture training. Feedback-based online training with two 45 min sessions per week can improve postural parameters in adolescent soccer players over a period of six weeks.

Author Contributions: Conceptualization, O.L., T.H., J.M. and T.S.; methodology, O.L.; software, C.D., T.H.; formal analysis, J.M., C.D. and O.L.; investigation, O.L., T.S. and J.M.; writing—original draft preparation, O.L., M.F. and C.D.; writing—review and editing, O.L., C.D. and M.F.; visualization, C.D and O.L. All authors have read and agreed to the published version of the manuscript.

Funding: This research received no external funding.

Institutional Review Board Statement: The study was conducted in accordance with the Declaration of Helsinki, and approved by the Institutional Ethics Committee (ref. nr. 6-18).

Informed Consent Statement: Informed consent was obtained from all subjects involved in the study. Written informed consent has been obtained from the subjects to publish this paper.

Data Availability Statement: The data is available for qualified requests.

Acknowledgments: The authors would like to thank Lena Reiter and Sebastian Eberle for conducting the exercises, Jonas Dully for his help in preparing the study and Diana Mergen for her assistance in analysing the posture photos.

Conflicts of Interest: The authors declare no conflict of interest.

References

1. Kratenova, J.; Zejglicova, K.; Malý, M.; Filipová, V. Prevalence and Risk Factors of Poor Posture in School Children in the Czech Republic. *J. Sch. Health* **2007**, *77*, 131–137. [CrossRef] [PubMed]
2. Wirth, B.; Humphreys, B.K. Pain characteristics of adolescent spinal pain. *BMC Pediatr.* **2015**, *15*, 42. [CrossRef] [PubMed]
3. Lee, J.-H. Effects of forward head posture on static and dynamic balance control. *J. Phys. Ther. Sci.* **2016**, *28*, 274–277. [CrossRef] [PubMed]
4. Czaprowski, D.; Stoliński, Ł.; Tyrakowski, M.; Kozinoga, M.; Kotwicki, T. Non-structural misalignments of body posture in the sagittal plane. *Scoliosis Spinal Disord.* **2018**, *13*, 6. [CrossRef] [PubMed]

5.   Dolphens, M.; Cagnie, B.; Coorevits, P. Sagittal standing posture and its association with spinal pain: A school-based epidemiological study of 1196 Flemish adolescents before age at peak height velocity. *Spine* **2012**, *37*, 1657–1666. [CrossRef]
6.   Kim, M.S.; Cha, Y.J.; Choi, J.D. Correlation between forward head posture, respiratory functions, and respiratory accessory muscles in young adults. *J. Back Musculoskelet. Rehabil.* **2017**, *30*, 711–715. [CrossRef]
7.   Buchtelová, E.; Tichy, M.; Vaniková, K. Influence of muscular imbalances on pelvic position and lumbar lordosis: A theoretical basis. *J. Nurs. Soc. Stud. Public Health Rehabil.* **2013**, *1*, 25–36.
8.   Ludwig, O.; Mazet, C.; Mazet, D.; Hammes, A.; Schmitt, E. Age-dependency of posture parameters in children and adolescents. *J. Phys. Ther. Sci.* **2016**, *28*, 1607–1610. [CrossRef] [PubMed]
9.   Foltran-Mescollotto, F.; Gonçalves, B.; de Castro-Carletti, E.M.; Oliveira, A.B.; Pelai, E.B.; Rodrigues-Bigaton, D. Smartphone addiction and the relationship with head and neck pain and electromiographic activity of masticatory muscles. *Work* **2021**, *68*, 633–640. [CrossRef] [PubMed]
10. David, D.; Giannini, C.; Chiarelli, F.; Mohn, A. Text neck syndrome in children and adolescents. *Int. J. Environ. Res. Public Health* **2021**, *18*, 1565. [CrossRef] [PubMed]
11. Schmidt, S.C.E.; Anedda, B.; Burchartz, A.; Eichsteller, A.; Kolb, S.; Nigg, C.; Niessner, C.; Oriwol, D.; Worth, A.; Woll, A. Physical activity and screen time of children and adolescents before and during the COVID-19 lockdown in Germany: A natural experiment. *Sci. Rep.* **2020**, *10*, 21780. [CrossRef] [PubMed]
12. Chambonniere, C.; Lambert, C.; Fearnbach, N.; Tardieu, M.; Fillon, A.; Genin, P.; Larras, B.; Melsens, P.; Bois, J.; Pereira, B.; et al. Effect of the COVID-19 lockdown on physical activity and sedentary behaviors in French children and adolescents: New results from the ONAPS national survey. *Eur. J. Integr. Med.* **2021**, *43*, 101308. [CrossRef] [PubMed]
13. Guan, H.; Okely, A.D.; Aguilar-Farias, N.; Del Pozo Cruz, B.; Draper, C.E.; El Hamdouchi, A.; Florindo, A.A.; Jáuregui, A.; Katzmarzyk, P.T.; Kontsevaya, A.; et al. Promoting healthy movement behaviours among children during the COVID-19 pandemic. *Lancet Child Adolesc. Health* **2020**, *4*, 416–418. [CrossRef]
14. Caneiro, J.P.; O'Sullivan, P.; Burnett, A.; Barach, A.; O'Neil, D.; Tveit, O.; Olafsdottir, K. The influence of different sitting postures on head/neck posture and muscle activity. *Man. Ther.* **2010**, *15*, 54–60. [CrossRef]
15. Ertekin, E.; Günaydın, Ö.E. Neck pain in rounded shoulder posture: Clinico-radiologic correlation by shear wave elastography. *Int. J. Clin. Pract.* **2021**, *75*, e14240. [CrossRef]
16. Ruivo, R.M.; Pezarat-Correia, P.; Carita, A.I. Cervical and shoulder postural assessment of adolescents between 15 and 17 years old and association with upper quadrant pain. *Braz. J. Phys. Ther.* **2014**, *18*, 364–371. [CrossRef]
17. Häkkinen, A.; Salo, P.; Tarvainen, U.; Wiren, K.; Ylinen, J. Effect of manual therapy and stretching on neck muscle strength and mobility in chronic neck pain. *J. Rehabil. Med.* **2007**, *39*, 575–579. [CrossRef]
18. Tjønndal, A. #Quarantineworkout: The Use of Digital Tools and Online Training Among Boxers and Boxing Coaches During the COVID-19 Pandemic. *Front. Sports Act. Living* **2020**, *2*, 589483. [CrossRef]
19. Parker, K.; Uddin, R.; Ridgers, N.D.; Brown, H.; Veitch, J.; Salmon, J.; Timperio, A.; Sahlqvist, S.; Cassar, S.; Toffoletti, K.; et al. The Use of Digital Platforms for Adults' and Adolescents' Physical Activity During the COVID-19 Pandemic (Our Life at Home): Survey Study. *J. Med. Internet Res.* **2021**, *23*, e23389. [CrossRef]
20. Rohleder, B. *Kinder- & Jugendstudie 2022 [Children & Youth Study 2022]*; Bitcom Research: Berlin, Germany, 2022; p. 4.
21. Fourchet, F.; Materne, O.; Rajeb, A.; Horobeanu, C.; Farooq, A. Pelvic Tilt: Reliability of Measuring the Standing Position and Range of Motion in Adolescent Athletes. *Br. J. Sports Med.* **2014**, *48*, 594. [CrossRef]
22. Hazar, Z.; Karabicak, G.O.; Tiftikci, U. Reliability of photographic posture analysis of adolescents. *J. Phys. Ther. Sci.* **2015**, *27*, 3123–3126. [CrossRef]
23. Sikka, I.; Chawla, C.; Seth, S.; Alghadir, A.H.; Khan, M. Effects of Deep Cervical Flexor Training on Forward Head Posture, Neck Pain, and Functional Status in Adolescents Using Computer Regularly. *BioMed Res. Int.* **2020**, *2020*, 8327565. [CrossRef]
24. Ludwig, O.; Fröhlich, M.; Schmitt, E. Therapy of poor posture in adolescents: Sensorimotor training increases the effectiveness of strength training to reduce increased anterior pelvic tilt. *Cogent Med.* **2016**, *3*, 1262094. [CrossRef]
25. Mair, P.; Wilcox, R. Robust statistical methods in R using the WRS2 package. *Behav. Res. Methods* **2020**, *52*, 464–488. [CrossRef]
26. Waskom, M.L. Seaborn: Statistical data visualization. *J. Open Source Softw.* **2021**, *6*, 3021. [CrossRef]
27. Dolphens, M.; Vansteelandt, S.; Cagnie, B.; Nijs, J.; Danneels, L. Factors associated with low back and neck pain in young adolescence: A multivariable modeling study. *Physiotherapy* **2015**, *101*, e1090–e1091. [CrossRef]
28. Ruivo, R.M.; Pezarat-Correia, P.; Carita, A.I. Effects of a Resistance and Stretching Training Program on Forward Head and Protracted Shoulder Posture in Adolescents. *J. Manip. Physiol. Ther.* **2017**, *40*, 1–10. [CrossRef]
29. Sheikhhoseini, R.; Shahrbanian, S.; Sayyadi, P.; O'Sullivan, K. Effectiveness of therapeutic exercise on forward head posture: A systematic review and meta-analysis. *J. Manip. Physiol. Ther.* **2018**, *41*, 530–539. [CrossRef]
30. Dolphens, M.; Vansteelandt, S.; Cagnie, B.; Vleeming, A.; Nijs, J.; Vanderstraeten, G.; Danneels, L. Multivariable modeling of factors associated with spinal pain in young adolescence. *Eur. Spine J.* **2016**, *25*, 2809–2821. [CrossRef]
31. Bruhn, S.; Kullmann, N.; Gollhofer, A. Combinatory Effects of High-Intensity-Strength Training and Sensorimotor Training on Muscle Strength. *Int. J. Sport. Med.* **2006**, *27*, 401–406. [CrossRef]
32. Bansal, S.; Katzman, W.B.; Giangregorio, L.M. Exercise for Improving Age-Related Hyperkyphotic Posture: A Systematic Review. *Arch. Phys. Med. Rehabil.* **2014**, *95*, 129–140.

33. Kim, D.; Cho, M.; Park, Y.; Yang, Y. Effect of an exercise program for posture correction on musculoskeletal pain. *J. Phys. Ther. Sci.* **2015**, *27*, 1791–1794. [CrossRef] [PubMed]
34. Park, H.C.; Kim, Y.-S.; Seok, S.-H.; Lee, S.-K. The effect of complex training on the children with all of the deformities including forward head, rounded shoulder posture, and lumbar lordosis. *J. Exerc. Rehabil.* **2014**, *10*, 172–175. [CrossRef] [PubMed]

45

Journal of
*Functional Morphology
and Kinesiology*

Article

The Impact of Nordic Walking Pole Length on Gait Kinematic Parameters

Luca Russo [1,*], Guido Belli [2], Andrea Di Blasio [3], Elena Lupu [4], Alin Larion [5], Francesco Fischetti [6], Eleonora Montagnani [7], Pierfrancesco Di Biase Arrivabene [8] and Marco De Angelis [8]

[1] Department of Human Sciences, Università Telematica degli Studi IUL, 50122 Florence, Italy
[2] Department of Sciences for Life Quality Studies, University of Bologna, 47921 Rimini, Italy
[3] Department of Medicine and Aging Sciences, "G. d'Annunzio" University of Chieti-Pescara, 66100 Chieti, Italy
[4] Department of Motor Activities, Petroleum Gas University Ploiesti, 100600 Ploiesti, Romania
[5] Faculty of Physical Education and Sport, Ovidius University of Constanta, 900029 Constanta, Romania
[6] Department of Basic Medical Sciences, Neuroscience and Sense Organs, University of Study of Bari, 70124 Bari, Italy
[7] Department of Sports and Health Sciences, University of Brighton, Brighton BN2 4AT, UK
[8] Department of Biotechnological and Applied Clinical Sciences, University of L'Aquila, 67100 L'Aquila, Italy
* Correspondence: l.russo@iuline.it

Abstract: Nordic walking (NW) is a popular physical activity used to manage chronic diseases and maintain overall health and fitness status. This study aimed to compare NW to ordinary walking (W) with regard to pole length and to identify kinematic differences associated with different poles' length (55%, 65% and 75% of the subject's height, respectively). Twelve male volunteers (21.1 ± 0.7 years; 1.74 ± 0.05 m; 68.9 ± 6.1 kg) were tested in four conditions (W, NW55, NW65 and NW75) at three different speeds ($4\text{-}5\text{-}6$ km$*$h^{-1}). Each subject performed a total of twelve tests in a random order. Three-dimensional kinematics of upper and lower body were measured for both W and NW, while oxygen consumption levels (VO_2) and rating of perceived exertion (RPE) were measured only for NW trials with different poles' length. NW showed a higher step length, lower elbow motion and higher trunk motion ($p < 0.05$) compared to W. Additionally, NW65 did not show any kinematic or RPE differences compared to NW55 and NW75. Only NW75 showed a higher elbow joint ($p < 0.05$) and lower pole ($p < 0.05$) range of motion compared to NW55 and a higher VO_2 ($p < 0.05$) compared to NW55 and NW65 at 6 km$*$h^{-1}. In conclusion, the use of the poles affects the motion of the upper and lower body during gait. Poles with shorter or longer length do not produce particular changes in NW kinematics. However, increasing the length of the pole can be a smart variation in NW to increase exercise metabolic demand without significantly affecting the kinematics and the RPE.

Keywords: Nordic walking; walking; 3D kinematics; biomechanics; gait analysis

Citation: Russo, L.; Belli, G.; Di Blasio, A.; Lupu, E.; Larion, A.; Fischetti, F.; Montagnani, E.; Di Biase Arrivabene, P.; De Angelis, M. The Impact of Nordic Walking Pole Length on Gait Kinematic Parameters. *J. Funct. Morphol. Kinesiol.* **2023**, *8*, 50. https://doi.org/10.3390/jfmk8020050

Academic Editor: Luca Petrigna

Received: 30 March 2023
Revised: 17 April 2023
Accepted: 23 April 2023
Published: 26 April 2023

1. Introduction

Walking with poles in an urban or natural environment is commonly known as Nordic walking (NW), defined as an outdoor non-competitive fitness activity in which the practitioners perform brisk walking with specially designed poles, similar to those for cross country skiing, engaging also the upper body [1–3]. NW is widely and successfully applied in the management of several not-transmissible chronic diseases, such as diabetes, cancer, hypertension, obesity [4–9] or in case of joints diseases [10,11] or for general health maintaining purpose [12,13].

Over an approximate period of 25 years, NW has received great attention by researchers. The main attractive feature of NW, as a fitness activity, is the suggestion from the literature about a higher request in terms of energy expenditure with respect to ordinary walking (W). The average rate of oxygen uptake (VO_2) is reported to be significantly higher in NW with respect to W at the same speed [3,14–21], without any increase in the rate of perceived exertion [14,17,19–21]. These results reinforce the practice of NW as an activity for weight loss and maintaining an optimal health status.

From a biomechanical point of view, NW is often compared to W with the aim to understand if the use of the poles could affect the joint load. For this reason, the lower limbs are mainly investigated [22], in particular the knee joint, obtaining, however, controversial results [2,23–26] due to the different tested samples in which not always the knee joint load was reduced in NW with respect to W. The upper body is mainly studied in terms of muscular activation patterns [27]. Some studies suggest that NW can lead to increased shoulder mobility and reduced tenderness in the shoulder girdle [28], as well as decreased neck and shoulder pain [29]. In general, NW increases the walking distance and speed, as well as the stride length, decreasing the cadence [30]. Moreover, NW increases the muscle activation and the strength of upper limbs, as well as the upper and lower range of motion and the ground reaction force [30].

According to the actual literature state of art, NW is deeply investigated but less attention is given to the main element that differentiates NW from W: the pole. In NW, the poles are investigated most of all for the force transmission profile [31,32] and it is well-known that properly pushing the pole on the ground allows to increase the NW metabolic benefits. Not by chance, there is a difference in the applied force on the ground between NW instructors and recreational individuals [32]. Despite that, at present, few information is available about the poles' length and how this parameter can affect the body kinematics. Usually, the NW poles are set approximately at 2/3 of the subject's height (height of the subject in centimeters∗0.68) [23,32,33], but it is common to use them based on comfort perception. For this reason, it is important to understand the number of kinematic modifications that different pole lengths could induce. From a metabolic perspective, some results are available about both different weight and length of the poles. About weight, no differences were measured comparing NW with heavier (adding 0.5; 1 and 1.5 kg to the pole) or standard poles on metabolic data, therefore there is no suggestion to use heavier poles to increase the metabolic demand [34]. About length, poles lengths which are self-selected caused greater energy expenditure than shorter poles only in uphill walking but not on level walking [3]. No information is present about using longer poles and no information is available about the kinematics of NW with different poles' length. Moreover, the increased metabolic demand of NW compared to W is well-known, but less is known about the relationships between kinematics and VO_2 consumption during NW exercise.

It is reasonable to hypothesize that modifying the length of NW poles could affect gait kinematics, as well as perceived fatigue and metabolic demand during exercise. Modifying the poles' length could lead to using the upper limbs in a different way to properly manage the pole, and this modification could affect the motion of the spine and lower limbs. Thus, this study aimed to (1) characterize the kinematics of NW compared to regular walking, with a focus on upper body motion and (2) examine the effects of different pole lengths on kinematics, metabolic demand and perceived fatigue during NW exercise. The primary practical goal of this research is to provide evidence-based recommendations for practitioners and instructors involved in NW training.

2. Materials and Methods

2.1. Design and Participants

This is a cross-sectional study design with a random crossover approach. Participants were recruited as part of the undergraduate program of Sport Science at the University of L'Aquila. The participation was on a voluntary base and the specific criteria of selection included: no history of musculoskeletal or neurological pain in the last 6 months and to be physically active. Participants never practiced NW. This was required as we wanted to resemble as close as possible the activity of recreational and novice NW practitioners; expert participants would have displayed different walking techniques and styles, biasing the results. Due to the inexperience of the participants, they attended three preliminary weeks of supervised training with a NW instructor, to avoid a learning effect during the study as Figard-Fabre et al. found [21]. The instructor taught the basic technique (diagonal technique) according to the INWA guidelines [35]. Diagonal technique can be briefly described as technique that seeks a position of the trunk inclined forward, with long strides and active use of the arms without any limitation of flexion or extension [36]. Practices were performed using a treadmill, at the same time of the day, 3 sessions/week, 1 h/session.

A total amount of 20 male participants were recruited but only 12 completed the experiment (mean ± SD: 21.1 ± 0.7 years; 1.74 ± 0.05 m; 68.9 ± 6.1 kg; 22.7 ± 1.6 BMI; 12.8 ± 2.7% body fat). A power analysis of the sample was conducted, and 12 subjects were sufficient to satisfy a power at 80% and an error probability at 5%. During the recruiting phase and prior to the training phase, all participants were informed and gave voluntary consent to participate in the study, and privacy criteria were meet. The study was approved by the Ovidius University of Constanta Nr. 141 din 21 February 2023 in accordance with the Declaration of Helsinki.

2.2. Instrumentation

The NW pole used was customized with a telescopic pole (Skitrab, Bormio, Italy; 210 g of mass each) connected with a grip and a tip specific for NW (Swix, Lillehammer, Norway). The choice to use a telescopic pole allowed to modify the length but not the weight of the pole for each trial.

Kinematic data were collected using a 4 infrared 3-dimensional camera SMART integrated System working at a sample rate of 60 Hz (BTS Bioengineering, Garbagnate Milanese, Italy.

Metabolic data were collected using a portable metabolimeter K4b2 (Cosmed, Roma, Italy), calibrated before each test session.

All the tests were performed on a Cosmed T170 treadmill (Cosmed, Roma, Italy), adapted by taking out frontal and lateral handlebars to allow participants to perform the typical NW gesture without hitting the structure. The treadmill was calibrated according to the literature [37] before the beginning of the experimental procedures.

The Figure 1 resumes the used instrumentation.

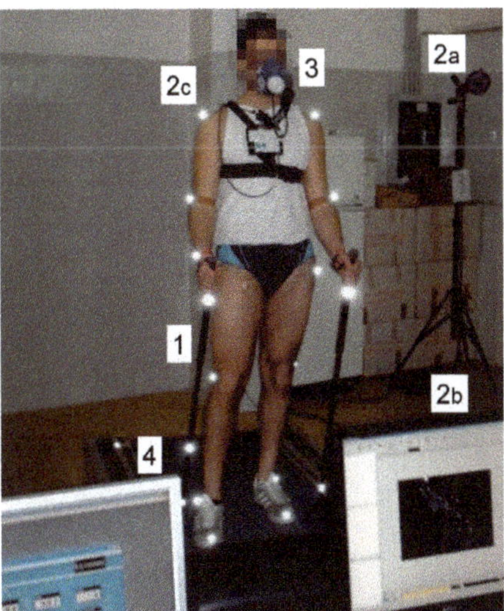

Figure 1. Schematic representation of the experimental setting. 1—Nordic walking pole. 2—3D kinematic integrated system; 2a—Infrared camera; 2b—Reflective marker; 2c—Workstation. 3—Metabolimeter. 4—Treadmill and workstation for treadmill management.

2.3. Procedure and Data Collection

Testing procedures were carried out in a Sport Performance Laboratory at a mean temperature of 19 °C and a mean relative humidity of 51%, and each subject was tested at the same time of the day, avoiding any circadian effect [38–40]. The first day of the training period, all participants were tested for anthropometric measures aiming to find out the three different poles' length: classical, reduced and increased length 65%, 55% and 75% of the participant's height, respectively. The supervised training period was performed with classical length poles (65% of subject's height) as well as longer and shorter poles in a random order. The last day of training, each individual received a personalized random sequence of the test conditions.

Each participant performed a total amount of 12 tests. The experimental procedure had 4 conditions, each condition was tested during a single day and each test day was separated by 7 days or rest. The test conditions were: ordinary walking (W) and NW with poles at different lengths (65%, 55%, 75% of the subject's height defined as NW65, NW55, NW75). The length of the pole was measured from the insertion of the lace in the handle to the tip final extremity (Figure 2). Each condition was tested at 3 different fixed speeds 4-5-6 km∗h^{-1} [41] and no slope was applied to the treadmill. Conditions were randomly assigned while speeds were always administered from the slowest to the faster one. Each trial lasted 10 min and a resting period of 20 min was observed between each trial test. Table 1 resumes a real example of randomized sequence for one single participant. During tests, kinematic data, rating of perceived exertion (RPE) and oxygen consumption (VO$_2$) were collected.

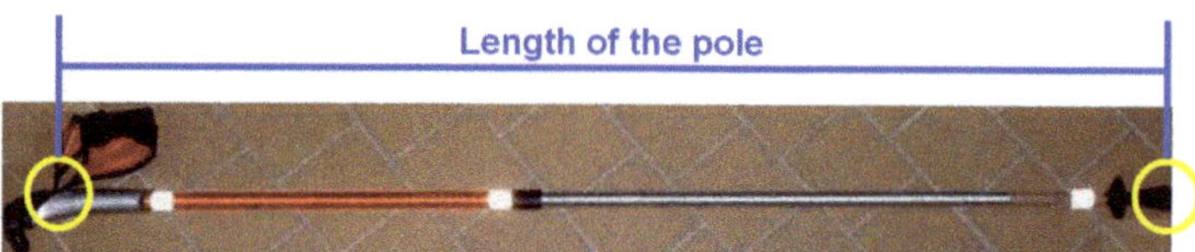

Figure 2. Landmarks for measurement of NW pole's length, from left to right: insertion of the lace in the handle and tip final extremity.

Table 1. Real tests sequence of a participant. The sequence of the conditions was randomly assigned while the speeds were always administered from the slowest to the faster one (4-5-6 km∗h^{-1}). Each trial, represented as an "X", lasted 10 min and a resting period of 20 min was observed between each trial test.

Condition	Week 1 Speed (km∗h^{-1})			Week 2 Speed (km∗h^{-1})			Week 3 Speed (km∗h^{-1})			Week 4 Speed (km∗h^{-1})		
	4	5	6	4	5	6	4	5	6	4	5	6
W							X	X	X			
NW55	X	X	X									
NW65										X	X	X
NW75				X	X	X						

W—Walking. NW55—Nordic walking with pole length adjusted at 55% of subject's height. NW65—Nordic walking with pole length adjusted at 65% of subject's height. NW75—Nordic walking with pole length adjusted at 75% of subject's height.

2.3.1. Kinematic Data

Kinematic analysis was conducted using a model of 18 body reflective markers (Figure 3), positioned on the most used body landmarks according to other studies about NW or motion analysis [2,23,25,26,42–47].

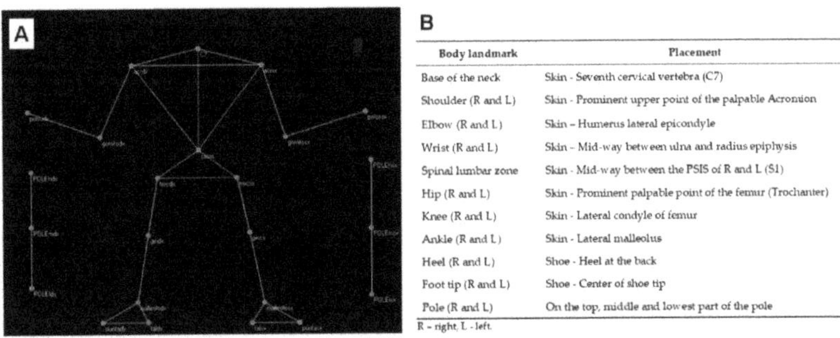

B

Body landmark	Placement
Base of the neck	Skin - Seventh cervical vertebra (C7)
Shoulder (R and L)	Skin - Prominent upper point of the palpable Acromion
Elbow (R and L)	Skin – Humerus lateral epicondyle
Wrist (R and L)	Skin – Mid-way between ulna and radius epiphysis
Spinal lumbar zone	Skin - Mid-way between the PSIS of R and L (S1)
Hip (R and L)	Skin - Prominent palpable point of the femur (Trochanter)
Knee (R and L)	Skin - Lateral condyle of femur
Ankle (R and L)	Skin - Lateral malleolus
Heel (R and L)	Shoe - Heel at the back
Foot tip (R and L)	Shoe - Center of shoe tip
Pole (R and L)	On the top, middle and lowest part of the pole

R – right, L - left.

Figure 3. Position of the reflective markers. (**A**)—Kinematic used model for the automatic recognition of the body landmarks. (**B**)—Detailed description of the position for each reflective marker on body landmarks.

The last two minutes of each trial were taken into account for the kinematic analysis, aiming to reduce the stride variability [48]. Smoothing procedures and missing frames were managed through the SMART Analyzer software (BTS Bioengineering, Garbagnate Milanese—Italy). Once the raw kinematic data were processed, a ratio between the step length and the elbow horizontal displacement was also calculated (SL/EHD ratio), with the aim to understand if the use of the poles addressed some changes in the upper limbs motion in NW with respect to W. All the kinematic data were measured on the right side of each subject in order to standardize the process. Table 2 resumes and describes the kinematic parameters acquired for each test.

Table 2. Detailed description of each parameter measured with kinematic analysis.

	Parameter	Description
Upper body	C7 vertical acceleration peak ($m*s^{-2}$)	Peak vertical acceleration after the push off phase of the foot on the ground
	C7 vertical displacement (cm)	Vertical displacement of the marker on C7 from the lowest to the highest point for each step
	Elbow horizontal displacement (cm)	Maximal horizontal displacement of the elbow on the sagittal plane for each step
	Elbow Δ angle (°)	Angular displacement of the elbow from the maximum flexion to the maximum extended position
	Elbow advancing speed ($m*s^{-1}$)	Speed of the elbow during the forward movement from back to front
	Spine forward slope (°)	Slope of the spine calculated by the inclination of the S1-C7 segment from the vertical position
	S1 vertical displacement (cm)	Vertical displacement of the marker on S1 from the lowest to the highest point for each step
Lower body	Step length (m)	Horizontal displacement of feet
	Step frequency (Hz)	Number of steps per second
Poles	Tip sagittal displacement (m)	Horizontal displacement of the pole tip
	Movement frequency (Hz)	Number of pole pushes per second
	Minimum slope (°)	Minimum inclination of the pole
	Maximum slope (°)	Maximum inclination of the pole
	Δ Slope (°)	Angular displacement of the pole from the maximum to the minimum inclination of the pole

Measures for elbows, feet and poles were taken all on the right side for the entire sample. Note: all the data are expressed as an average value of all gait cycles.

2.3.2. Fatigue and Metabolic Data

VO_2 was measured using a wearable metabolimeter during the whole duration of each test and the last seven minutes were taken into account for the metabolic analysis. The evaluation of the average oxygen consumption of each NW trial was used to look for correlations between kinematics and metabolic demand in NW exercise. Due to the well-known metabolic difference between NW and W [3,14–21], VO_2 consumption during W was not considered in this study. Anyway, the metabolimeter was worn as well even during W trials in order to avoid perceived comfort difference with respect to NW. Immediately after each trial, the participants indicated their rating of perceived exertion using the category rating-10 (CR-10) scale modified by Foster et al. [49].

2.4. Statistical Analysis

All data were tested with the Shapiro–Wilk's test for normality. As data were normally distributed, parametric inferences were used for the analysis. Kinematic differences between NW65 and W were tested using the Student paired *t*-test, while to measure kinematics differences in NW with different poles' height, an ANOVA design for repeated measures with Sidak correction was used. Finally, the Pearson correlation coefficient was used to look for a correlation between kinematic and metabolic data. The significant level was set at $p = 0.05$ and the data were analyzed using SPSS (SPSS Inc., Chicago, IL, USA).

3. Results

3.1. Kinematic Differecences between NW65 and W

According to statistical analysis, kinematic data in NW65 showed significant differences with respect to W, for all the three tested speeds. Most of the significant differences involved the upper body segments (Table 3).

Table 3. Differences between NW65 and W for kinematics.

Parameter	4 km*h^{-1} NW Mean (SD)	4 km*h^{-1} W Mean (SD)	*p* Value	5 km*h^{-1} NW Mean (SD)	5 km*h^{-1} W Mean (SD)	*p* Value	6 km*h^{-1} NW Mean (SD)	6 km*h^{-1} W Mean (SD)	*p* Value
C7 vertical acceleration peak (m*s^{-2})	3.0 (0.3)	3.0 (0.3)	0.805	3.3 (0.3)	3.1 (0.3)	0.127	3.7 (0.4)	3.5 (0.4)	0.002 *
C7 vertical displacement (cm)	3.5 (0.5)	3.2 (0.5)	0.010 *	4.7 (0.7)	4.0 (0.7)	0.000 *	5.7 (1.0)	5.1 (0.8)	0.007 *
Elbow horizontal displacement (cm)	10.9 (4.3)	15.9 (2.9)	0.002 *	12.7 (3.4)	17.4 (2.9)	0.001 *	14.1 (4.3)	18.3 (3.3)	0.001 *
Elbow Δ angle (°)	34.0 (8.6)	22.1 (6.5)	0.005 *	35.3 (8.8)	29.8 (5.7)	0.141	36.4 (10.3)	36.5 (6.9)	0.970
Elbow advancing speed (m*s^{-1})	0.2 (0.1)	0.3 (0.1)	0.003 *	0.2 (0.1)	0.3 (0.1)	0.001 *	0.3 (0.1)	0.4 (0.1)	0.000 *
Spine forward slope (°)	9.3 (2.6)	8.7 (2.1)	0.109	10.0 (3.3)	9.1 (2.3)	0.032 *	10.9 (2.5)	10.1 (2.1)	0.067
S1 vertical displacement (cm)	3.1 (0.6)	2.7 (0.6)	0.001 *	4.5 (1.5)	3.7 (0.8)	0.033 *	5.5 (1.1)	4.9 (0.9)	0.020 *
Step length (m)	0.66 (0.04)	0.64 (0.03)	0.036 *	0.74 (0.04)	0.71 (0.04)	0.000 *	0.80 (0.05)	0.77 (0.04)	0.006 *
Step frequency (Hz)	0.85 (0.08)	0.86 (0.03)	0.750	0.91 (0.04)	0.95 (0.03)	0.004 *	0.99 (0.05)	1.02 (0.03)	0.010 *
SL/EHD ratio (cm)	7.3 (4.1)	4.12 (0.8)	0.023 *	6.3 (2.0)	4.2 (0.7)	0.004 *	6.4 (2.5)	4.4 (0.8)	0.005 *

W—walking. NW65—Nordic walking with pole length adjusted at 65% of subject's height. SL/EHD ratio—ratio between the step length and the elbow horizontal displacement. * significant difference between NW and W.

The use of the poles significantly increased both the C7 and the S1 vertical displacement for all speeds; even the elbow horizontal displacement and the elbow advancing speed resulted significantly lower in NW compared to W, while the step length was higher in NW in comparison with W for all tested speeds.

3.2. Kinematic Differecences between NW with Different Poles' Length

A limited number of significant differences were found for kinematics between NW with different poles' length (Figure 4). For all the three tested speeds, no differences were measured between NW65 and NW75, while only one significant difference was measured at 6 km*h^{-1} between NW55 and NW65 for the pole Δ slope (31.7 ± 2.6° and 28.0 ± 3.2°, respectively, $p = 0.013$). Significant differences were measured between the shorter and the longer poles' length, NW55 and NW75, particularly on the elbow and poles kinematics. At 4 km*h^{-1}, three significant differences were measured between NW55 and NW75: (1) pole minimum slope (14.5 ± 6.6° and 25.7 ± 10.7°, respectively, *p* = 0.013); (2) pole maximum slope (41.8 ± 4.4° and 49.5 ± 4.0°, respectively, *p* = 0.028); (3) pole Δ slope (27.3 ± 3.2° and 23.4 ± 2.5°, respectively, *p* = 0.010). At 5 km∗h^{-1}, four significant differences were measured between NW55 and NW75: (1) elbow Δ angle (30.7 ± 6.5° and 41.9 ± 13.3°, respectively, *p* = 0.027); (2) pole minimum slope (13.5 ± 5.5° and 27.4 ± 9.0°, respectively, *p* = 0.000); (3) pole maximum slope (42.9 ± 4.0° and 51.1 ± 7.1°, respectively, *p* = 0.001); (4) pole Δ slope (29.4 ± 2.7° and 24.7 ± 2.8°, respectively, *p* = 0.003). Finally, at 6 km∗h^{-1}, four significant differences were measured between NW55 and NW75: (1) elbow Δ angle (31.7 ± 7.1° and 42.8 ± 12.0°, respectively, *p* = 0.031); (2) pole minimum slope (13.7 ± 6.4° and 27.7 ± 7.0°, respectively, *p* = 0.000); (3) pole maximum slope (45.5 ± 4.3° and 54.2 ± 5.3°, respectively, *p* = 0.002); (4) pole Δ slope (31.7 ± 2.6° and 26.5 ± 3.0°, respectively, *p* = 0.000). The results for all the data measured can be read in Appendix A.

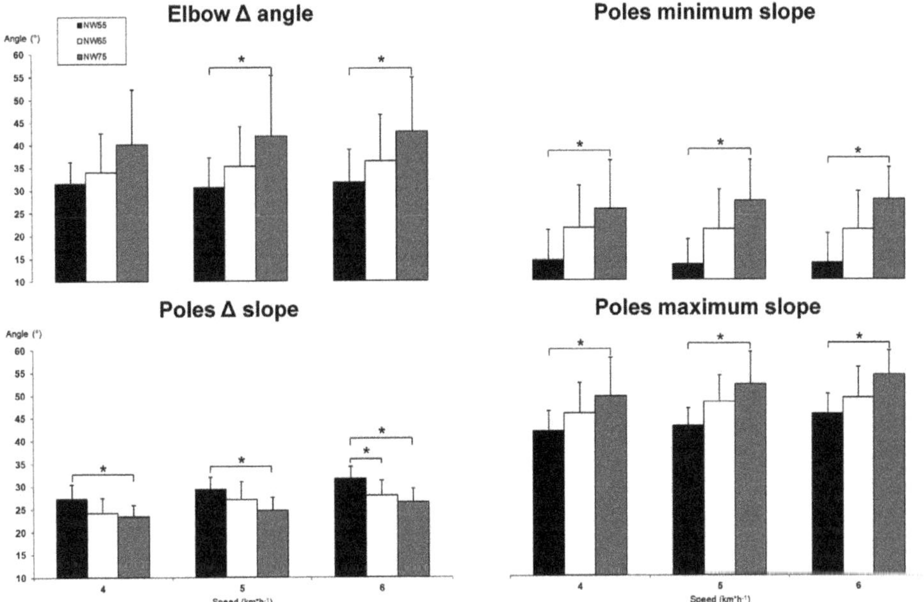

Figure 4. Kinematic differences between NW using different poles' length. NW55—Nordic walking with pole length adjusted at 55% of subject's height. NW65—Nordic walking with pole length adjusted at 65% of subject's height. NW75—Nordic walking with pole length adjusted at 75% of subject's height. * significant difference.

3.3. Correlations between Metabolic Data and Kinematics in NW

No differences were measured in RPE for NW with different poles' lengths at the same speed; while two significant differences were measured at 6 km∗h^{-1} between NW55 and NW75, as well as between NW65 and NW75 for VO$_2$ (21.7 ± 1.7 mL∗kg^{-1}∗min^{-1} and 24.1 ± 2.2 mL∗kg^{-1}∗min^{-1}, respectively, *p* = 0.002 and 22.2 ± 2.9 mL∗kg^{-1}∗min^{-1} and 24.1 ± 2.2 mL∗kg^{-1}∗min^{-1}, respectively, *p* = 0.030) (Figure 5).

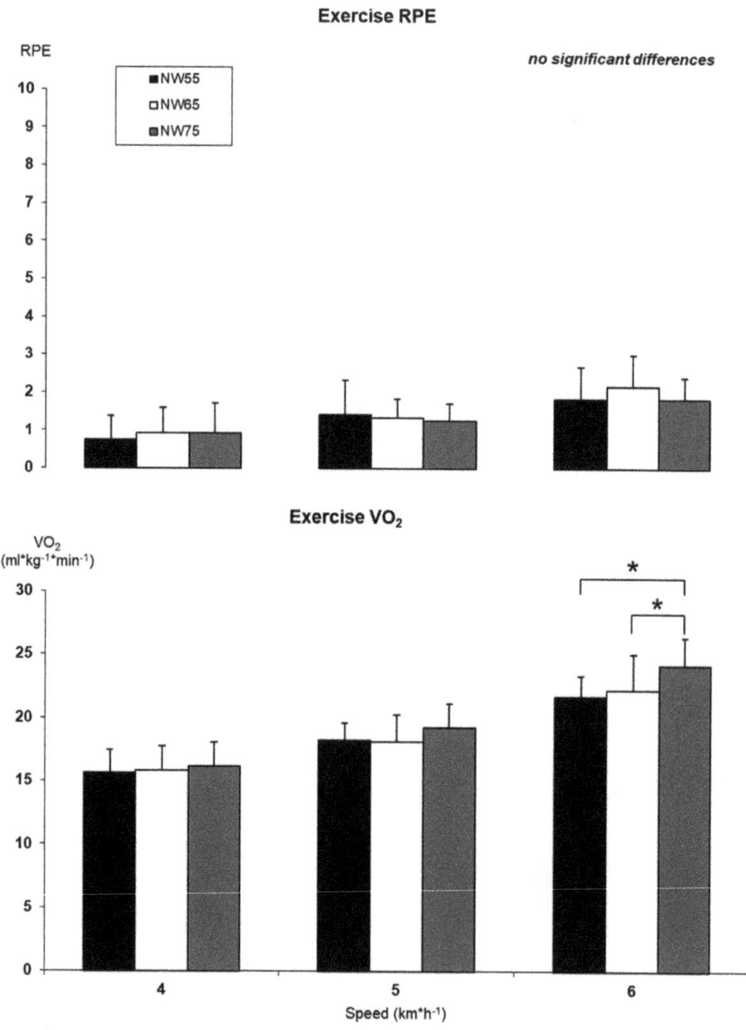

Figure 5. Fatigue perception and oxygen consumption differences between NW using different poles' length. RPE—Rating of perceived exertion. VO$_2$—Oxygen consumption. NW55—Nordic walking with pole length adjusted at 55% of subject's height. NW65—Nordic walking with pole length adjusted at 65% of subject's height. NW75—Nordic walking with pole length adjusted at 75% of subject's height. * significant difference.

Significant correlations were measured between metabolic data and kinematics in NW. In NW65 at 6 km∗h^{-1}, the S1 vertical displacement significantly correlated with VO$_2$ (r = 0.59 p = 0.043). In NW75 at 4, 5 and 6 km∗h^{-1}, significant correlations were found between the spine forward slope and the VO$_2$ (r = 0.71 p = 0.010; r = 0.69 p = 0.014; r = 0.71 p = 0.010, respectively, for each tested speed). In NW75, another two correlations were measured: one at 4 km∗h^{-1} between poles Δ slope and VO$_2$ (r = 0.63 p = 0.027) and another one at 6 km∗h^{-1} between elbow Δ angle and VO$_2$ (r = 0.61 p = 0.036). No significant correlations were found at any speed for NW55.

4. Discussion

This study aimed to characterize the kinematics of NW compared to W and to examine the effects of different pole lengths on kinematics, metabolic demand and perceived fatigue during NW. The study of the effect of different poles' length on gait kinematics is the main novelty of the investigation. Additionally, exploring the relationship between NW kinematics and exercise VO_2 consumption could provide a new perspective on NW training and practice.

The main attention in NW from a biomechanical point of view is usually focused on lower limbs [22]. Conversely, the most interesting aspect of this investigation is the use of a 3D kinematic model for the study of the upper body segments. Finally, by utilizing a treadmill, we were able to maintain a constant speed throughout the tests, resulting in a more accurate comparison of the impact of various pole lengths on both kinematic and metabolic data.

4.1. Kinematic Differecences between NW65 and W

Significant differences in kinematics were found for all the tested speeds between NW65 and W. The use of the poles during walking seems to affect the gait kinematics both for the upper and lower body. The measured differences are significant but relatively small. This point can be considered delusive, but it should be considered that the NW is a physical activity that is performed for a long time during the training session, therefore it is logical to mind that even small differences can affect the final energy expenditure of the exercise.

The poles affect the vertical displacement of the two segments of the spine, C7 and S1, increasing the muscular work against gravity. In fact, in NW65, a moderate positive but significant correlation is present between the vertical displacement of S1 and the VO_2. The use of the poles in NW not only allows a higher use of the upper limbs musculature [50] but engages all the trunk muscles [51–53], reflecting a higher vertical displacement of the whole spine.

With respect to the upper limb motion, particular attention was paid to the elbow. A significant difference for the angle displacement of the elbow was measured between NW and W only for the slower speed, 4 km$*$h^{-1}, in fact NW showed a higher angle displacement with respect to W. This difference disappeared when the gait speed increases, higher speeds showed a similar elbow angle. On the other hand, significant differences were observed between NW and W at all tested speeds, both in terms of elbow horizontal displacement and elbow advancing speed. In both cases, the NW values were smaller with respect to W, this can be addressed by the presence of the poles. It is reasonable to hypothesize that the pole's presence can affect the motion of the upper limbs both on the coordinative and on the kinematics profile, due to the necessity to push the pole against the ground [32]. It is well-known that the pole's weight does not affect the oxygen consumption, but only the muscle activation [34]; thus, the motion of the elbow is mainly affected by the pole's usage but not by its weight. The result of the present study cannot confirm at all this hypothesis but the presence of the positive significant correlation for NW75 between the elbow angle displacement and the VO_2 can suggest that more research in this area is needed to clear this aspect.

For the lower body kinematics, NW showed a significantly higher step length with respect to W and this aspect could explain the advantages of NW for special populations, such as diabetic people [54,55]. To increase the step length results in engaging the ankle and hip motion much more, therefore this information can be relevant in all cases where higher ankle and hip motion is required: such as sedentary or pathological individuals [10,11,56]. The results of this research, about step length and step frequency, are consistent with previous studies [19,25] but with different absolute values. Willson et al. [25] observed an increase in step length up to 20 cm, whereas the difference observed in the present research is smaller (approximately 3 cm). However, it is important to note that there are methodological differences between the two experimental protocols as other researchers used self-selected and faster walking speeds compared to the present study. Therefore, it is reasonable that the difference in step length could depend on the absolute different speed used for the tests.

Finally, as a direct consequence of the effect of the poles on the elbow motion and step length, even the SL/EHD ratio is significantly different between NW and W. The SL/EHD ratio is the mathematical ratio between the step length and the elbow horizontal displacement. It could be considered as a coordinative aspect of the gait. For all tested speeds, NW showed higher values of this ratio compared to W, because NW led to higher step length but smaller elbow horizontal displacement. This information confirms that the use of the poles can modify the gait spatiotemporal parameters.

4.2. Characterization of NW with Different Poles' Length

The second aim of this research was to examine the effects of different pole lengths on kinematics, metabolic demand and perceived fatigue during NW. Usually, the suggested length of the poles is set approximately at 2/3 of the subject's height [23,32,33]; thus, in this research, the poles were set at 55%, 65% and 75% of the subject's height. It would have been plausible to hypothesize that the use of shorter or longer poles could have influenced the spine inclination or the upper limb motion with respect to the classical length, but no statistical kinematic differences were measured between NW65 and NW55 or NW75 for all the tested speeds. No statistical differences were measured even for exercise RPE, suggesting that the length of the poles does not affect the perceived fatigue at any speed.

The use of different poles' length determined a limited number of significant differences both on the kinematic and metabolic profile: all the differences were related to the NW75 condition. In fact, NW75 showed a higher elbow angular displacement and a lower change of poles' slope with respect to the ground compared to NW55. The significant difference was present for all tested speeds. It is rationale to assume that the higher elbow angular displacement and the position of the pole constantly inclined with respect to the ground enable the participants to optimize the use of the pole during the pushing phase. The latter interpretation can be read in light of the significant increase of VO_2 consumption at 6 km$*$h^{-1} measured in NW75 compared to NW55 and NW65. These results are consistent with literature that suggests keeping the pole inclined for pushing optimization [21]. Therefore, the use of a longer pole can be beneficial for NW practitioners due to the increased metabolic demand without any effect on kinematics and exercise RPE compared to classical length. On the other hand, no reasons seem to be present for suggesting to reduce the length of the pole (NW55) in level walking, whereas it seems useful on uphill walking [3].

The absence of substantial significant kinematic differences between the different pole length conditions may seem odd, but it is important to consider that the sample for this research was deliberately chosen to have no previous experience. Although the participants underwent a training period before the tests, they were not as skilled as regular practitioners. Therefore, it is plausible to assume that the absence of specific NW coordination and technique allowed the participants to respond in a similar way to the different pole lengths. This perspective can be viewed positively when working with novice subjects, but this data cannot be applied to skilled practitioners without specific research.

Based on the research findings, NW instructors can opt for a different pole length compared to the traditional one when instructing novices. This is because it is widely known that the level of comfort does not vary between poles with different lengths [3]. Furthermore, novice practitioners may be advised by NW instructors to use a longer pole length (75% of the subject's height) if their objective is to increase metabolic demand during exercise without altering the kinematics of the movement and the exercise RPE. This recommendation may be particularly useful in the management of metabolic clinical conditions, such as overweight and obesity [55,57,58]. This practical application can be considered the main novelty of this research.

4.3. Limitations

The findings of this study are applicable solely to young male novice NW practitioners. It would be highly interesting to replicate the same experimental protocol using participants of different genders and ages. Additionally, this research did not provide any insights into the subject's adaptation or learning process during the training period, nor did it examine the comfort levels associated with the use of a specific pole during extended training sessions. Therefore, this study may be viewed as a preliminary exploration of the topic, and further research is warranted to enhance the practical information available to NW instructors.

5. Conclusions

Nordic walking (NW) differs from ordinary walking in both kinematic and metabolic profiles. The use of poles influences the movement of the upper and lower body. The length of the poles, whether shorter or longer than the standard size (about 2/3 of the subject's height), does not significantly alter the kinematics of NW, with the exception of a comparison between longer and shorter poles, where longer poles significantly impact elbow angular displacement and pole angle relative to the ground. These differences may explain the higher VO_2 consumption associated with longer poles, without any corresponding increase in exercise RPE. In summary, this study sheds light on pole-length selection, suggesting that reducing the length of poles offers no benefit to ground-level NW, while increasing pole length provides a viable method to increase metabolic demand without impacting kinematics or perceived fatigue. Furthermore, the use of telescoping poles may offer a convenient way to adapt poles to varying needs.

Author Contributions: Conceptualization, L.R. and M.D.A.; methodology, M.D.A.; software, L.R. and P.D.B.A.; validation, L.R.; formal analysis, A.D.B.; investigation, L.R. and P.D.B.A.; resources, L.R.; data curation, L.R.; writing—original draft preparation, G.B.; writing—review and editing, L.R.; visualization, F.F. and E.M.; supervision, M.D.A.; project administration, E.L. and A.L. All authors have read and agreed to the published version of the manuscript.

Funding: This research received no external funding.

Institutional Review Board Statement: The study was conducted in accordance with the Declaration of Helsinki and approved by the Ethics Committee of OVIDIUS UNIVERISTY OF CONSTANTA (protocol code No. 141 din 21 February 2023).

Informed Consent Statement: Informed consent was obtained from all subjects involved in the study.

Data Availability Statement: The data that support the findings of this study are available from the corresponding author, upon reasonable request.

Conflicts of Interest: The authors declare no conflict of interest.

Appendix A

The Appendix A contains:

- All the figures for the kinematic parameters described in Table 3;
- All the results for the data described into the Section 3.2. Kinematic differences between NW with different poles' length.

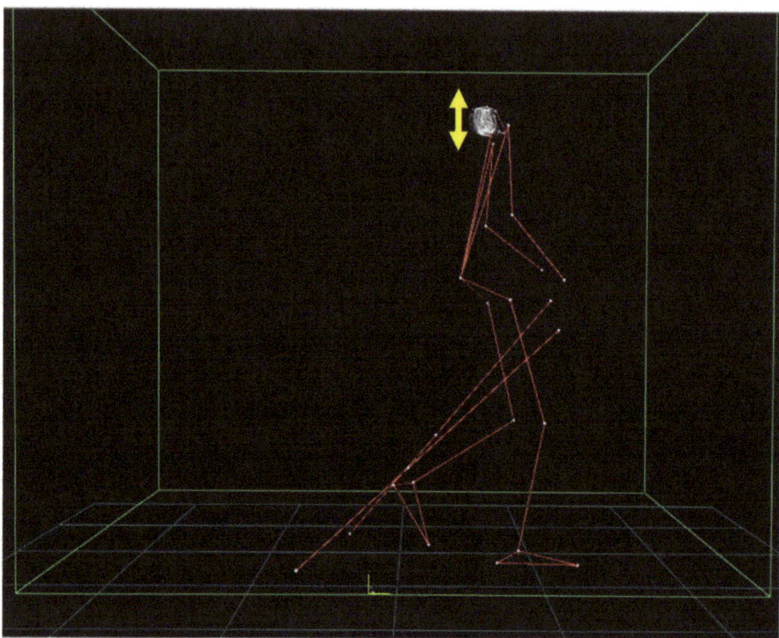

Figure A1. C7 vertical displacement (used to calculate acceleration).

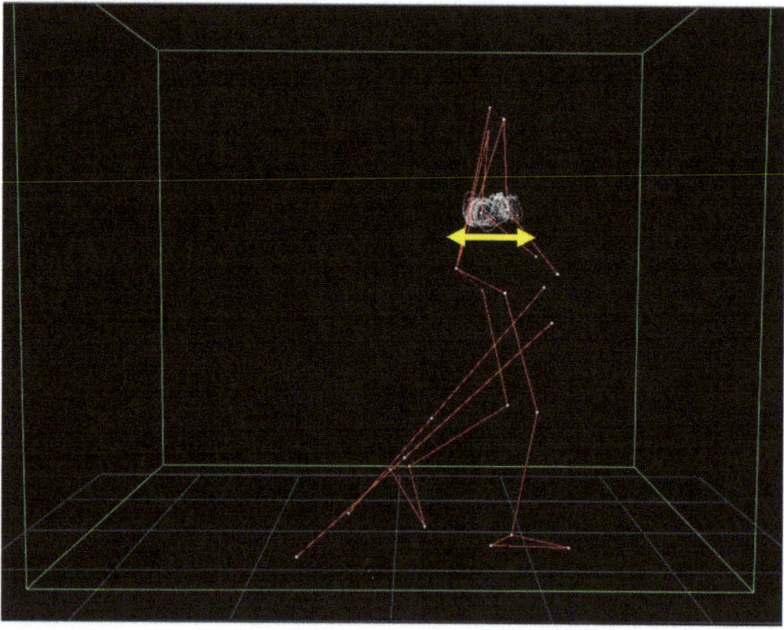

Figure A2. Elbow horizontal displacement (used to calculate advancing speed).

Figure A3. Elbow Δ angle (maximal extension and maximal flexion).

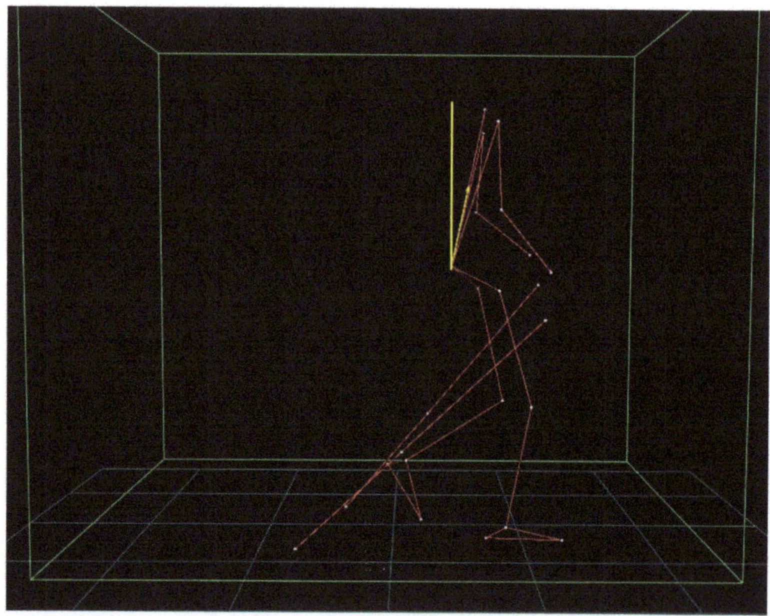

Figure A4. Spine forward slope.

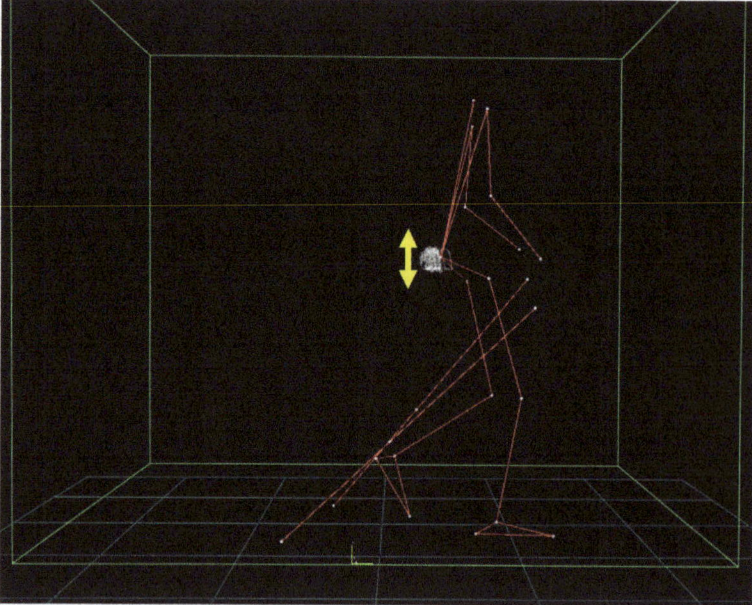

Figure A5. S1 vertical displacement.

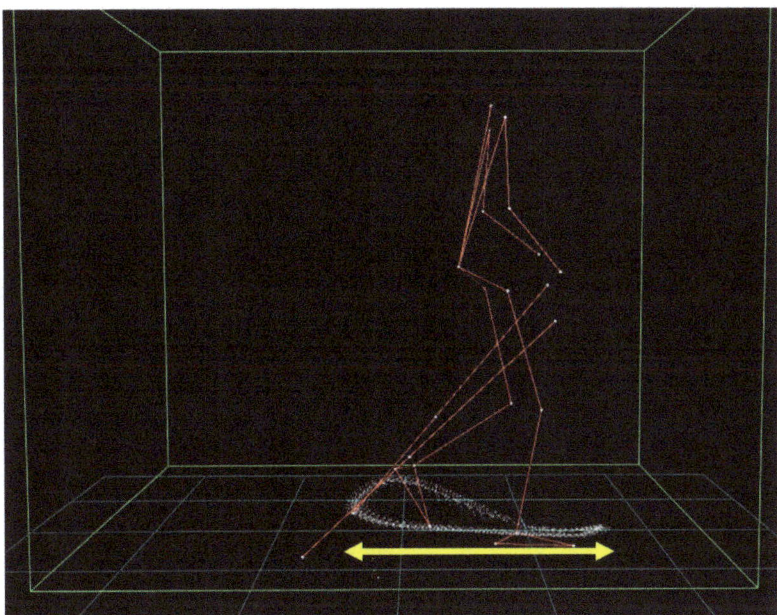

Figure A6. Step length.

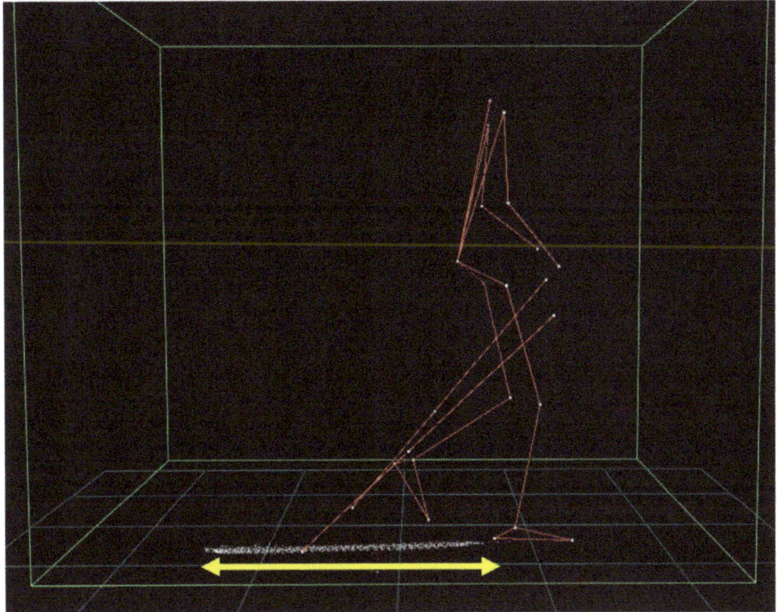

Figure A7. Tip sagittal displacement.

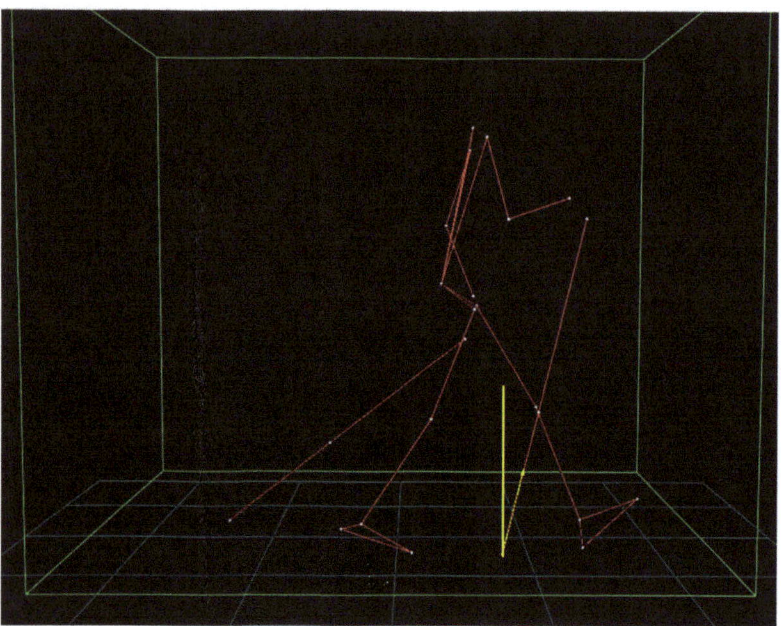

Figure A8. Pole minimum slope.

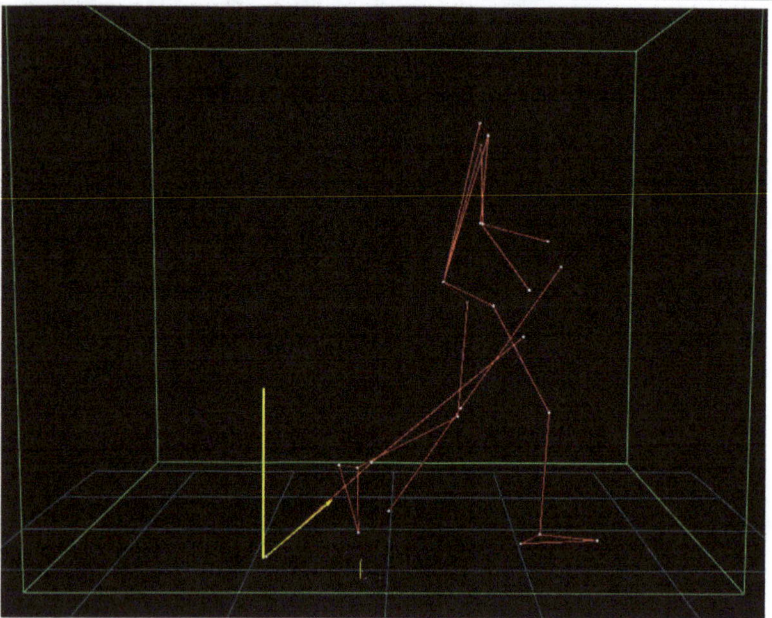

Figure A9. Pole maximum slope.

Table A1. Kinematic differences between NW55, NW65 and NW75 at 4 km∗h^{-1}.

Parameter	4 km∗h^{-1}					
	NW55 Mean (SD)	**NW65** Mean (SD)	**NW75** Mean (SD)	*p* Value 55–65	*p* Value 65–75	*p* Value 55–75
C7 vertical acceleration peak (m∗s^{-2})	3.1 (0.2)	3.0 (0.3)	3.1 (0.3)	0.948	0.894	0.998
C7 vertical displacement (cm)	3.7 (0.4)	3.5 (0.5)	3.8 (0.5)	0.559	0.322	0.973
Elbow horizontal displacement (cm)	10.9 (3.0)	10.9 (4.3)	12.2 (4.5)	1.000	1.000	0.802
Elbow Δ angle (°)	31.6 (4.7)	34.0 (8.6)	40.1 (12.2)	0.888	0.293	0.081
Elbow advancing speed (m∗s^{-1})	0.2 (0.1)	0.2 (0.1)	0.2 (0.1)	0.948	0.855	0.549
Spine forward slope (°)	9.7 (3.1)	9.3 (2.6)	9.3 (2.6)	0.968	1.000	0.982
S1 vertical displacement (cm)	3.2 (0.5)	3.1 (0.6)	3.1 (0.8)	0.946	1.000	0.946
Step length (m)	0.68 (0.03)	0.66 (0.04)	0.67 (0.03)	0.590	0.963	0.857
Step frequency (Hz)	0.82 (0.05)	0.85 (0.08)	0.82 (0.04)	0.520	0.412	0.998
Tip sagittal displacement (m)	0.63 (0.04)	0.62 (0.03)	0.64 (0.03)	0.913	0.480	0.853
Movement frequency (Hz)	0.82 (0.04)	0.83 (0.04)	0.82 (0.04)	0.748	1.000	0.763
Minimum slope (°)	14.5 (6.6)	21.6 (9.3)	25.7 (10.7)	0.176	0.606	0.013 *
Maximum slope (°)	41.8 (4.4)	45.8 (6.7)	49.5 (4.0)	0.411	0.475	0.028 *
Δ Slope (°)	27.3 (3.2)	24.2 (3.2)	23.4 (2.5)	0.054	0.884	0.010 *
SL/EHD ratio (cm)	6.8 (2.6)	7.3 (4.1)	6.4 (3.1)	0.979	0.879	0.985

NW55, Nordic walking with pole length adjusted at 55% of subject's height. NW65, Nordic walking with pole length adjusted at 65% of subject's height. NW75, Nordic walking with pole length adjusted at 75% of subject's height. * significant difference.

Table A2. Kinematic differences between NW55, NW65 and NW75 at 5 km∗h^{-1}.

Parameter	5 km∗h^{-1}					
	NW55 Mean (SD)	**NW65** Mean (SD)	**NW75** Mean (SD)	*p* Value 55–65	*p* Value 65–75	*p* Value 55–75
C7 vertical acceleration peak (m∗s^{-2})	3.2 (0.3)	3.3 (0.3)	3.3 (0.4)	1.000	0.962	0.949
C7 vertical displacement (cm)	4.7 (0.6)	4.7 (0.7)	4.6 (0.8)	0.990	0.999	0.967
Elbow horizontal displacement (cm)	13.0 (3.2)	12.7 (3.4)	13.7 (4.0)	0.993	0.870	0.960
Elbow Δ angle (°)	30.7 (6.5)	35.3 (8.8)	41.9 (13.3)	0.599	0.298	0.027 *
Elbow advancing speed (m∗s^{-1})	0.2 (0.1)	0.2 (0.1)	0.2 (0.1)	1.000	0.558	0.536

Table A2. *Cont.*

Parameter	5 km$*$h^{-1}					
	NW55 Mean (SD)	NW65 Mean (SD)	NW75 Mean (SD)	*p* Value 55–65	*p* Value 65–75	*p* Value 55–75
Spine forward slope (°)	10.1 (3.1)	10.0 (3.3)	10.0 (2.7)	1.000	1.000	1.000
S1 vertical displacement (cm)	4.4 (0.7)	4.5 (1.5)	4.0 (1.1)	0.994	0.761	0.885
Step length (m)	0.74 (0.03)	0.74 (0.04)	0.7 (0.03)	0.987	0.721	0.522
Step frequency (Hz)	0.91 (0.03)	0.91 (0.04)	0.93 (0.04)	0.944	0.848	0.532
Tip sagittal displacement (m)	0.68 (0.04)	0.69 (0.05)	0.69 (0.03)	0.906	0.950	0.999
Movement frequency (Hz)	0.91 (0.03)	0.91 (0.04)	0.93 (0.04)	0.971	0.499	0.759
Minimum slope (°)	13.5 (5.5)	21.1 (8.8)	27.4 (9.0)	0.071	0.171	0.000 *
Maximum slope (°)	42.9 (4.0)	48.2 (5.8)	52.1 (7.1)	0.090	0.289	0.001 *
Δ Slope (°)	29.4 (2.7)	27.1 (3.9)	24.7 (2.8)	0.231	0.213	0.003 *
SL/EHD ratio (cm)	6.0 (1.4)	6.3 (2.0)	5.8 (2.1)	0.972	0.900	0.994

NW55, Nordic walking with pole length adjusted at 55% of subject's height. NW65, Nordic walking with pole length adjusted at 65% of subject's height. NW75, Nordic walking with pole length adjusted at 75% of subject's height. * significant difference.

Table A3. Kinematic differences between NW55, NW65 and NW75 at 6 km$*$h^{-1}.

Parameter	6 km$*$h^{-1}					
	NW55 Mean (SD)	NW65 Mean (SD)	NW75 Mean (SD)	*p* Value 55–65	*p* Value 65–75	*p* Value 55–75
C7 vertical acceleration peak (m$*$s^{-2})	3.7 (0.4)	3.7 (0.4)	3.7 (0.5)	0.996	0.997	0.973
C7 vertical displacement (cm)	5.8 (0.8)	5.7 (1.0)	5.7 (1.2)	0.992	0.999	0.973
Elbow horizontal displacement (cm)	14.2 (3.7)	14.1 (4.3)	15.6 (4.4)	1.000	0.752	0.811
Elbow Δ angle (°)	31.7 (7.21)	36.4 (10.2)	42.8 (12.0)	0.602	0.328	0.031 *
Elbow advancing speed (m$*$s^{-1})	0.3 (0.1)	0.3 (0.1)	0.3 (0.1)	1.000	0.685	0.644
Spine forward slope (°)	11.4 (2.9)	10.9 (2.5)	11.3 (2.8)	0.969	0.985	1.000
S1 vertical displacement (cm)	5.65 (0.8)	5.49 (1.1)	5.35 (1.6)	0.989	0.970	0.869
Step length (m)	0.81 (0.04)	0.80 (0.05)	0.79 (0.04)	0.995	0.822	0.691
Step frequency (Hz)	0.98 (0.03)	0.99 (0.05)	1.00 (0.04)	0.929	0.966	0.711

Table A3. *Cont.*

Parameter	NW55 Mean (SD)	NW65 Mean (SD)	NW75 Mean (SD)	p Value 55–65	p Value 65–75	p Value 55–75
			$6 \text{ km} * \text{h}^{-1}$			
Tip sagittal displacement (m)	0.73 (0.04)	0.74 (0.06)	0.74 (0.06)	0.913	0.998	0.964
Movement frequency (Hz)	0.99 (0.04)	0.99 (0.05)	1.00 (0.04)	1.000	0.963	0.975
Minimum slope (°)	13.7 (6.4)	21.1 (8.4)	27.7 (7.0)	0.059	0.095	0.000 *
Maximum slope (°)	45.5 (4.3)	49.1 (6.8)	54.2 (5.3)	0.337	0.088	0.002 *
Δ Slope (°)	31.7 (2.6)	28.0 (3.2)	26.5 (3.0)	0.013 *	0.500	0.000 *
SL/EHD ratio (cm)	6.00 (1.4)	6.38 (2.6)	5.47 (1.7)	0.943	0.579	0.884

NW55, Nordic walking with pole length adjusted at 55% of subject's height. NW65, Nordic walking with pole length adjusted at 65% of subject's height. NW75, Nordic walking with pole length adjusted at 75% of subject's height. * significant difference.

References

1. Kukkonen-Harjula, K.; Hiilloskorpi, H.; Mänttäri, A.; Pasanen, M.; Parkkari, J.; Suni, J.; Fogelholm, M.; Laukkanen, R. Self-Guided Brisk Walking Training with or without Poles: A Randomized-Controlled Trial in Middle-Aged Women. *Scand. J. Med. Sci. Sport.* **2007**, *17*, 316–323. [CrossRef] [PubMed]
2. Hansen, L.; Henriksen, M.; Larsen, P.; Alkjaer, T. Nordic Walking Does Not Reduce the Loading of the Knee Joint. *Scand. J. Med. Sci. Sport.* **2008**, *18*, 436–441. [CrossRef] [PubMed]
3. Hansen, E.A.; Smith, G. Energy Expenditure and Comfort during Nordic Walking with Different Pole Lengths. *J. Strength Cond. Res.* **2009**, *23*, 1187–1194. [CrossRef] [PubMed]
4. Tschentscher, M.; Niederseer, D.; Niebauer, J. Health Benefits of Nordic Walking: A Systematic Review. *Am. J. Prev. Med.* **2013**, *44*, 76–84. [CrossRef] [PubMed]
5. Fischer, M.J.; Krol-Warmerdam, E.M.M.; Ranke, G.M.C.; Vermeulen, H.M.; Van Der Heijden, J.; Nortier, J.W.R.; Kaptein, A.A. Stick Together: A Nordic Walking Group Intervention for Breast Cancer Survivors. *J. Psychosoc. Oncol.* **2015**, *33*, 278–296. [CrossRef]
6. Sánchez-Lastra, M.A.; Torres, J.; Martínez-Lemos, I.; Ayán, C. Nordic Walking for Women with Breast Cancer: A Systematic Review. *Eur. J. Cancer Care Engl.* **2019**, *28*, e13130. [CrossRef]
7. Della Guardia, L.; Carnevale Pellino, V.; Filipas, L.; Bonato, M.; Gallo, G.; Lovecchio, N.; Vandoni, M.; Codella, R. Nordic Walking Improves Cardiometabolic Parameters, Fitness Performance, and Quality of Life in Older Adults With Type 2 Diabetes. *Endocr. Pract.* **2023**, *29*, 135–140. [CrossRef]
8. Gram, B.; Christensen, R.; Christiansen, C.; Gram, J. Effects of Nordic Walking and Exercise in Type 2 Diabetes Mellitus: A Randomized Controlled Trial. *Clin. J. Sport Med.* **2010**, *20*, 355–361.
9. Di Blasio, A.; Morano, T.; Napolitano, G.; Bucci, I.; Di Santo, S.; Gallina, S.; Cugusi, L.; Di Donato, F.; D'Arielli, A.; Cianchetti, E. Nordic Walking and the Isa Method for Breast Cancer Survivors: Effects on Upper Limb Circumferences and Total Body Extracellular Water—A Pilot Study. *Breast Care Basel* **2016**, *11*, 428–431. [CrossRef]
10. Bieler, T.; Siersma, V.; Magnusson, S.P.; Kjaer, M.; Christensen, H.E.; Beyer, N. In Hip Osteoarthritis, Nordic Walking Is Superior to Strength Training and Home-Based Exercise for Improving Function. *Scand. J. Med. Sci. Sport.* **2017**, *27*, 873–886. [CrossRef]
11. O'Donovan, R.; Kennedy, N. "Four Legs Instead of Two"—Perspectives on a Nordic Walking-Based Walking Programme among People with Arthritis. *Disabil. Rehabil.* **2015**, *37*, 1635–1642. [CrossRef]
12. Grigoletto, A.; Mauro, M.; Oppio, A.; Greco, G.; Fischetti, F.; Cataldi, S.; Toselli, S. Effects of Nordic Walking Training on Anthropometric, Body Composition and Functional Parameters in the Middle-Aged Population. *Int. J. Environ. Res. Public Health* **2022**, *19*, 7433. [CrossRef]
13. Bullo, V.; Gobbo, S.; Vendramin, B.; Duregon, F.; Cugusi, L.; Di Blasio, A.; Bocalini, D.S.; Zaccaria, M.; Bergamin, M.; Ermolao, A. Nordic Walking Can Be Incorporated in the Exercise Prescription to Increase Aerobic Capacity, Strength, and Quality of Life for Elderly: A Systematic Review and Meta-Analysis. *Rejuvenation Res.* **2018**, *21*, 141–161. [CrossRef]
14. Rodgers, C.D.; VanHeest, J.L.; Schachter, C.L. Energy expenditure during submaximal walking with Exerstriders. *Med. Sci. Sport. Exerc.* **1995**, *27*, 607–611. [CrossRef]
15. Walter, P.R.; Porcari, J.P.; Brice, G.; Terry, L. Acute Responses to Using Walking Poles in Patients with Coronary Artery Disease. *J. Cardiopulm. Rehabil.* **1996**, *16*, 245–250. [CrossRef]

16. Porcari, J.P.; Hendrickson, T.L.; Walter, P.R.; Terry, L.; Walsko, G. The Physiological Responses to Walking with and without Power Poles on Treadmill Exercise. *Res. Q. Exerc. Sport* **1997**, *68*, 161–166. [CrossRef]
17. Church, T.S.; Earnest, C.P.; Morss, G.M. Field Testing of Physiological Responses Associated with Nordic Walking. *Res. Q. Exerc. Sport* **2002**, *73*, 296–300. [CrossRef]
18. Schiffer, T.; Knicker, A.; Hoffman, U.; Harwig, B.; Hollmann, W.; Strüder, H.K. Physiological Responses to Nordic Walking, Walking and Jogging. *Eur. J. Appl. Physiol.* **2006**, *98*, 56–61. [CrossRef]
19. Perrey, S.; Fabre, N. Exertion during uphill, level and downhill walking with and without hiking poles. *J. Sport. Sci. Med.* **2008**, *7*, 32–38.
20. Saunders, M.J.; Hipp, G.R.; Wenos, D.L.; Deaton, M.L. Trekking Poles Increase Physiological Responses to Hiking without Increased Perceived Exertion. *J. Strength Cond. Res.* **2008**, *22*, 1468–1474. [CrossRef]
21. Figard-Fabre, H.; Fabre, N.; Leonardi, A.; Schena, F. Physiological and Perceptual Responses to Nordic Walking in Obese Middle-Aged Women in Comparison with the Normal Walk. *Eur. J. Appl. Physiol.* **2010**, *108*, 1141–1151. [CrossRef] [PubMed]
22. Dziuba, A.K.; Żurek, G.; Garrard, I.; Wierzbicka-Damska, I. Biomechanical Parameters in Lower Limbs during Natural Walking and Nordic Walking at Different Speeds. *Acta Bioeng. Biomech.* **2015**, *17*, 95–101. [CrossRef] [PubMed]
23. Schwameder, H.; Roithner, R.; Müller, E.; Niessen, W.; Raschner, C. Knee Joint Forces during Downhill Walking with Hiking Poles. *J. Sport. Sci.* **1999**, *17*, 969–978. [CrossRef]
24. Jensen, S.B.; Henriksen, M.; Aaboe, J.; Hansen, L.; Simonsen, E.B.; Alkjær, T. Is It Possible to Reduce the Knee Joint Compression Force during Level Walking with Hiking Poles? *Scand. J. Med. Sci. Sport.* **2011**, *21*, e195–e200. [CrossRef] [PubMed]
25. Willson, J.; Torry, M.R.; Decker, M.J.; Kernozek, T.; Steadman, J.R. Effects of Walking Poles on Lower Extremity Gait Mechanics. *Med. Sci. Sport. Exerc.* **2001**, *33*, 142–147. [CrossRef] [PubMed]
26. Stief, F.; Kleindienst, F.I.; Wiemeyer, J.; Wedel, F.; Campe, S.; Krabbe, B. Inverse Dynamic Analysis of the Lower Extremities during Nordic Walking, Walking, and Running. *J. Appl. Biomech.* **2008**, *24*, 351–359. [CrossRef]
27. Pellegrini, B.; Peyré-Tartaruga, L.A.; Zoppirolli, C.; Bortolan, L.; Bacchi, E.; Figard-Fabre, H.; Schena, F. Exploring Muscle Activation during Nordic Walking: A Comparison between Conventional and Uphill Walking. *PLoS ONE* **2015**, *10*, e0138906. [CrossRef]
28. Kocur, P.; Pospieszna, B.; Choszczewski, D.; Michalowski, L.; Wiernicka, M.; Lewandowski, J. The Effects of Nordic Walking Training on Selected Upper-Body Muscle Groups in Female-Office Workers: A Randomized Trial. *Work* **2017**, *56*, 277–283. [CrossRef]
29. Saeterbakken, A.H.; Nordengen, S.; Andersen, V.; Fimland, M.S. Nordic Walking and Specific Strength Training for Neck- and Shoulder Pain in Office Workers: A Pilot-Study. *Eur. J. Phys. Rehabil. Med.* **2017**, *53*, 928–935. [CrossRef]
30. Roy, M.; Grattard, V.; Dinet, C.; Soares, A.V.; Decavel, P.; Sagawa, Y.J. Nordic Walking Influence on Biomechanical Parameters: A Systematic Review. *Eur. J. Phys. Rehabil. Med.* **2020**, *56*, 607–615. [CrossRef]
31. Schiffer, T.; Knicker, A.; Dannöhl, R.; Strüder, H.K. Energy Cost and Pole Forces during Nordic Walking under Different Surface Conditions. *Med. Sci. Sport. Exerc.* **2009**, *41*, 663–668. [CrossRef]
32. Fujita, E.; Yakushi, K.; Takeda, M.; Islam, M.M.; Nakagaichi, M.; Taaffe, D.R.; Takeshima, N. Proficiency in Pole Handling during Nordic Walking Influences Exercise Effectiveness in Middle-Aged and Older Adults. *PLoS ONE* **2018**, *13*, e0208070. [CrossRef]
33. Park, S.K.; Yang, D.J.; Kang, Y.H.; Kim, J.H.; Uhm, Y.H.; Lee, Y.S. Effects of Nordic Walking and Walking on Spatiotemporal Gait Parameters and Ground Reaction Force. *J. Phys. Ther. Sci.* **2015**, *27*, 2891–2893. [CrossRef]
34. Schiffer, T.; Knicker, A.; Montanarella, M.; Strüder, H.K. Mechanical and Physiological Effects of Varying Pole Weights during Nordic Walking Compared to Walking. *Eur. J. Appl. Physiol.* **2011**, *111*, 1121–1126. [CrossRef]
35. What Is Nordic Walking | INWA. Available online: https://www.inwa-nordicwalking.com/about-us/what-is-nordic-walking/ (accessed on 30 March 2023).
36. Encarnación-Martínez, A.; Catalá-Vilaplana, I.; Aparicio, I.; Sanchis-Sanchis, R.; Priego-Quesada, J.I.; Jimenez-Perez, I.; Pérez-Soriano, P. Does Nordic Walking technique influence the ground reaction forces? *Gait Posture* **2023**, *101*, 35–40. [CrossRef]
37. Padulo, J.; Chamari, K.; Ardigò, L.P. Walking and Running on Treadmill: The Standard Criteria for Kinematics Studies. *Muscles Ligaments Tendons J.* **2014**, *4*, 159. [CrossRef]
38. Ardigò, L.P.; Buglione, A.; Russo, L.; Cular, D.; Esposito, F.; Doria, C.; Padulo, J. Marathon Shoes vs. Track Spikes: A Crossover Pilot Study on Metabolic Demand at Different Speeds in Experienced Runners. *Res. Sport. Med.* **2023**, *31*, 13–20. [CrossRef]
39. Russo, L.; Montagnani, E.; Pietrantuono, D.; D'Angona, F.; Fratini, T.; Di Giminiani, R.; Palermi, S.; Ceccarini, F.; Migliaccio, G.M.; Lupu, E.; et al. Self-Myofascial Release of the Foot Plantar Surface: The Effects of a Single Exercise Session on the Posterior Muscular Chain Flexibility after One Hour. *Int. J. Environ. Res. Public Health* **2023**, *20*, 974. [CrossRef]
40. Ardigò, L.P.; Palermi, S.; Padulo, J.; Dhahbi, W.; Russo, L.; Linetti, S.; Cular, D.; Tomljanovic, M. External Responsiveness of the SuperOp™ Device to Assess Recovery After Exercise: A Pilot Study. *Front. Sport. Act. Living* **2020**, *2*, 67. [CrossRef]
41. Gomeñuka, N.A.; Oliveira, H.B.; Silva, E.S.; Costa, R.R.; Kanitz, A.C.; Liedtke, G.V.; Schuch, F.B.; Peyré-Tartaruga, L.A. Effects of Nordic walking training on quality of life, balance and functional mobility in elderly: A randomized clinical trial. *PLoS ONE* **2019**, *14*, e0211472. [CrossRef]
42. Leardini, A.; Sawacha, Z.; Paolini, G.; Ingrosso, S.; Nativo, R.; Benedetti, M.G. A New Anatomically Based Protocol for Gait Analysis in Children. *Gait Posture* **2007**, *26*, 560–571. [CrossRef] [PubMed]

43. Sibella, F.; Frosio, I.; Schena, F.; Borghese, N.A. 3D Analysis of the Body Center of Mass in Rock Climbing. *Hum. Mov. Sci.* **2007**, *26*, 841–852. [CrossRef] [PubMed]
44. Ferrari, A.; Benedetti, M.G.; Pavan, E.; Frigo, C.; Bettinelli, D.; Rabuffetti, M.; Crenna, P.; Leardini, A. Quantitative Comparison of Five Current Protocols in Gait Analysis. *Gait Posture* **2008**, *28*, 207–216. [CrossRef] [PubMed]
45. McGinley, J.L.; Baker, R.; Wolfe, R.; Morris, M.E. The Reliability of Three-Dimensional Kinematic Gait Measurements: A Systematic Review. *Gait Posture* **2009**, *29*, 360–369. [CrossRef]
46. Manca, M.; Leardini, A.; Cavazza, S.; Ferraresi, G.; Marchi, P.; Zanaga, E.; Benedetti, M.G. Repeatability of a New Protocol for Gait Analysis in Adult Subjects. *Gait Posture* **2010**, *32*, 282–284. [CrossRef]
47. Di Giminiani, R.; Di Lorenzo, D.; La Greca, S.; Russo, L.; Masedu, F.; Totaro, R.; Padua, E. Angle-Angle Diagrams in the Assessment of Locomotion in Persons with Multiple Sclerosis: A Preliminary Study. *Appl. Sci.* **2022**, *12*, 7223. [CrossRef]
48. Russo, L.; Di Capua, R.; Arnone, B.; Borrelli, M.; Coppola, R.; Esposito, F.; Padulo, J. Shoes and Insoles: The Influence on Motor Tasks Related to Walking Gait Variability and Stability. *Int. J. Environ. Res. Public Health* **2020**, *17*, 4569. [CrossRef]
49. Foster, C.; Florhaug, J.A.; Franklin, J.; Gottschall, L.; Hrovatin, L.A.; Parker, S.; Doleshal, P.; Dodge, C. A new approach to monitoring exercise training. *J. Strength Cond. Res.* **2001**, *15*, 109–115.
50. Boccia, G.; Zoppirolli, C.; Bortolan, L.; Schena, F.; Pellegrini, B. Shared and Task-Specific Muscle Synergies of Nordic Walking and Conventional Walking. *Scand. J. Med. Sci. Sport.* **2018**, *28*, 905–918. [CrossRef]
51. Hanuszkiewicz, J.M.; Woźniewski, M.; Malicka, I. The Influence of Nordic Walking on Isokinetic Trunk Muscle Endurance and Sagittal Spinal Curvatures in Women after Breast Cancer Treatment. *Acta Bioeng. Biomech.* **2020**, *22*, 47–54. [CrossRef]
52. Hanuszkiewicz, J.; Woźniewski, M.; Malicka, I. The Influence of Nordic Walking on Isokinetic Trunk Muscle Endurance and Sagittal Spinal Curvatures in Women after Breast Cancer Treatment: Age-Specific Indicators. *Int. J. Environ. Res. Public Health* **2021**, *18*, 2409. [CrossRef]
53. Peyré-Tartaruga, L.A.; Boccia, G.; Feijó Martins, V.; Zoppirolli, C.; Bortolan, L.; Pellegrini, B. Margins of Stability and Trunk Coordination during Nordic Walking. *J. Biomech.* **2022**, *134*, 111001. [CrossRef]
54. Pérez-Soriano, P.; Llana-Belloch, S.; Encarnación-Martínez, A.; Martínez-Nova, A.; Morey-Klapsing, G. Nordic Walking Practice Might Improve Plantar Pressure Distribution. *Res. Q. Exerc. Sport* **2011**, *82*, 593–599. [CrossRef]
55. Pippi, R.; Di Blasio, A.; Aiello, C.; Fanelli, C.; Bullo, V.; Gobbo, S.; Cugusi, L.; Bergamin, M. Effects of a Supervised Nordic Walking Program on Obese Adults with and without Type 2 Diabetes: The C.U.R.I.A.Mo. Centre Experience. *J. Funct. Morphol. Kinesiol.* **2020**, *5*, 62. [CrossRef]
56. Cugusi, L.; Solla, P.; Serpe, R.; Carzedda, T.; Piras, L.; Oggianu, M.; Gabba, S.; Di Blasio, A.; Bergamin, M.; Cannas, A.; et al. Effects of a Nordic Walking Program on Motor and Non-Motor Symptoms, Functional Performance and Body Composition in Patients with Parkinson's Disease. *NeuroRehabilitation* **2015**, *37*, 245–254. [CrossRef]
57. Gobbo, S.; Bullo, V.; Roma, E.; Duregon, F.; Bocalini, D.S.; Rica, R.L.; Di Blasio, A.; Cugusi, L.; Vendramin, B.; Bergamo, M.; et al. Nordic Walking Promoted Weight Loss in Overweight and Obese People: A Systematic Review for Future Exercise Prescription. *J. Funct. Morphol. Kinesiol.* **2019**, *4*, 36. [CrossRef]
58. Di Blasio, A.; Rinaldi, M.; Morano, T.; Izzicupo, P.; Dell'Aquila, S.; Grossi, S.; Russo, L.; Bergamin, M.; Gobbo, S.; Viscioni, G.; et al. Effects of Acupuncture and Nordic Walking Practice, and Their Interaction, on Bodily Fluids Distribution of Breast Cancer Survivors. *Hum. Mov.* **2022**, *24*, 130–139. [CrossRef]

Journal of
*Functional Morphology
and Kinesiology*

Article

Evaluation of Influencing Factors on the Maximum Climbing Specific Holding Time: An Inferential Statistics and Machine Learning Approach

Carlo Dindorf *, Eva Bartaguiz, Jonas Dully, Max Sprenger, Anna Merk, Stephan Becker, Michael Fröhlich and Oliver Ludwig

Department of Sports Science, Technische Universität Kaiserslautern, 67663 Kaiserslautern, Germany
* Correspondence: carlo.dindorf@sowi.uni-kl.de; Tel.: +49-(0)-631-205-5172

Abstract: Handgrip strength (HGS) appears to be an indicator of climbing performance. The transferability of HGS measurements obtained using a hand dynamometer and factors that influence the maximal climbing-specific holding time (CSHT) are largely unclear. Forty-eight healthy subjects (27 female, 21 male; age: 22.46 ± 3.17 years; height: 172.76 ± 8.91 cm; weight: 69.07 ± 12.41 kg; body fat: 20.05% ± 7.95%) underwent a maximal pull-up test prior to the experiment and completed a self-assessment using a Likert scale questionnaire. HGS was measured using a hand dynamometer, whereas CSHT was measured using a fingerboard. Multiple linear regressions showed that weight, maximal number of pull-ups, HGS normalized by subject weight, and length of the middle finger had a significant effect on the maximal CSHT (non-dominant hand: R^2_{corr} = 0.63; dominant hand: R^2_{corr} = 0.55). Deeper exploration using a machine learning model including all available data showed a predictive performance with R^2 = 0.51 and identified another relevant parameter for the regression model. These results call into question the use of hand dynamometers and highlight the performance-related importance of body weight in climbing practice. The results provide initial indications that finger length may be used as a sub-factor in talent scouting.

Keywords: climbing; exhaustion; fatigue; training; machine learning; sports; gender; training; data mining; artificial intelligence

Citation: Dindorf, C.; Bartaguiz, E.; Dully, J.; Sprenger, M.; Merk, A.; Becker, S.; Fröhlich, M.; Ludwig, O. Evaluation of Influencing Factors on the Maximum Climbing Specific Holding Time: An Inferential Statistics and Machine Learning Approach. *J. Funct. Morphol. Kinesiol.* **2022**, *7*, 95. https://doi.org/10.3390/jfmk7040095

Academic Editor: Luca Petrigna

Received: 2 October 2022
Accepted: 25 October 2022
Published: 27 October 2022

Publisher's Note: MDPI stays neutral with regard to jurisdictional claims in published maps and institutional affiliations.

1. Introduction

Handgrip strength (HGS) appears to be a crucial factor in climbing success [1]. Although HGS has a major impact on climbing performance, it has also been shown to play the biggest role in bouldering, followed by lead climbing, then speed climbing [2]. Nevertheless, Ref. [3] criticizes the transferability of HGS, which is usually measured using a hand dynamometer and represents static and isometric force production, which is not quite similar to climbing-specific determinants, e.g., the maximal climbing-specific holding time (CSHT) [4]. In the field of general HGS research, measurements are commonly performed using hand dynamometers, and factors such as finger length [5], body weight [6] and constitution [7–9], hand volume [10], as well as age and gender [6,11] have been identified as possible determining factors. This could be an indication that these also determine the maximal climbing-specific holding time; however, empirical verification is still pending as only a few studies have investigated factors affecting CSHT [12].

Knowledge about factors influencing CSHT could be of relevance because these factors could have implications for better training methods or performance assessments. Therefore, in addition to classic inferential statistical approaches, machine learning modeling could further show the predictive power of respective parameters and be beneficial in terms of identifying new relevant parameters (see, e.g., [13,14]). To the best of our knowledge, there are currently only a few studies that have used ML methods in the context of sports climbing (e.g., in the classification and generation of routes for MoonBoard climbing [15]).

Therefore, the actual potential of this methodology in the context of climbing seems to be largely unexamined.

Therefore, in this study, we analyzed whether there was a relation between different parameters (i.e., gender, weight, finger length, HGS, body fat, and maximal number of pull-ups) and maximal CSHT; we then tried to find a model that could predict the CSHT as accurately as possible using machine learning algorithms.

2. Materials and Methods

2.1. Subjects and Data Acquisition

Initially, data was collected from 52 subjects. For four subjects, the test was prematurely terminated because of their higher risk of injury as they were not able to hold their upper body in a stable position when hanging on the fingerboard. The data of these individuals were excluded from further analysis. Furthermore, illness and recent injuries in the upper extremities were exclusion criteria. For the final analysis, data from 48 healthy subjects were used (sex: 27 women, 21 men; age: 22.46 ± 3.17 years; height: 172.76 ± 8.91 cm; weight: 69.07 ± 12.41 kg; body fat: 20.05% ± 7.95%). Ten subjects had previous experience in sports climbing and trained one to three times a week without being active in sports climbing competitions. The other test participants were recreational athletes outside of climbing as well as people with little physical activity. The study was approved by the ethical committee of the university and met the criteria of the Declaration of Helsinki (World Medical Association, 2013). All participants signed informed consent forms, including permission to publish the results of the study. Subjects were asked not to perform intense physical activities for 48 h prior to the study and to come in a rested state.

2.2. General Measurement Procedure

Prior to the study, body composition was measured using a bioelectric impedance scale (InBody 770, InBody Europe, Eschborn, Germany). Height, upper and lower arm perimeter, and finger lengths were manually measured. Furthermore, the test subjects were asked for written information about their dominant (dH) and non-dominant (ndH) hand.

Additionally, using a 5-point Likert scale (0 = severely underdeveloped; 4 = well above average), the subjects were asked to what extent the following statements were true: (a) "My grip strength is heavily used in my everyday activities." (b) "My grip strength is heavily used in my private (e.g., sporting) activities." In addition, using an 11-point Likert scale (0 = severely underdeveloped; 10 = well above average), they were asked to provide subjective assessments of their CSHT.

Afterward, the maximum number of pull-ups was determined for each participant using a straight pull-up bar with radioulnar pronation of the palms. Correct execution criteria were chin at grip height at the upper position and full extension of the arms in the lower position. The grip width was set to shoulder width. Participants rested for at least 15 min before further measurements were taken after the pull-ups.

After the pull-ups, HGS was assessed using the Jamar hydraulic hand dynamometer (JLW Instruments, Chicago, IL, USA). Two measures of HGS were taken from each hand. We started with the dominant hand. Then, the non-dominant side was measured, before second measurements were carried out for each side. Following the method of [16], the maximum value achieved with either hand was used as the participant's isometric HGS. For the determination of the reference values, measurements were carried out according to the instructions in [17]. Again, participants rested for at least 15 min before further measurements were taken after the HGS measurements.

The maximal CSHT was determined by measuring the maximal time without touching the ground on a four-centimeter-deep crimp with rounded edges and structuring (characteristics corresponded to typical climbing holds used in the gym) of a MOON fingerboard (Moon Climbing Limited, Sheffield, UK). Test participants maintained a dead hang position with straight arms with their feet lifted about 20 cm from the ground. The holds were cleaned before each run. Loose magnesium was directly applied to the fingers before carrying out the measurement, as this is performed in sports practice to improve friction [18]. The holding was performed without the use of the thumb, with three or four fingers, as habitually preferred by the subjects. An image of this is displayed in Figure 1. Directly after each termination (fall-off event), subjects were asked for subjective ratings of their perceived exertion using the OMNI scale [19].

Figure 1. Hand positioning on the fingerboard.

2.3. Further Data Preparation and Calculations

Linear regression models for the modeling of CSHT were separately applied for the the dH and ndH variables to compare the results with each other. Backward elimination was applied to reduce the number of initial input variables. Dummy coding was performed for the categorical variable for sex (0 = woman, 1 = man). To consider possible multicollinearity, the following initial input variables were selected: sex, weight, body fat, maximal number of pull-ups, HGS normalized by the subject's weight (dH and ndH), length of the middle finger (dH and ndH), and length of the little finger (dH and ndH). These were based on background knowledge derived from previous studies [5–7,11], as they seemed to have a high impact on HGS or CSHT. No auto-correlations were present according to Durbin–Watson statistics. Further necessary requirements were checked and could be assumed. Statistical analysis was performed using IBM SPSS (version 26, SPSS Inc., Chicago, IL, USA).

Modeling of the CSHT using machine learning was performed to evaluate the predictive performance of the measured variables, as well as further evaluate the selected variables in the final regression model. Therefore, elastic net regression was applied using the standard hyperparameters of Scikit-learn [20]. The elastic net model was especially appropriate when highly correlated features were present [21], as was the case with the available data. Explorative modeling was performed using all measured variables as input features (see Table 1).

Table 1. Overview of the variables used as input features. dH = dominant hand; ndH = non-dominant hand.

Variable	Description
Gender Height Weight Muscle mass Body fat Lean leg mass	General anthropometric measurements
HGS dH HGS ndH	Hand grip strength (HGS) measured using hand dynamometer
HGS dH/kg HGS ndH/kg	HGS values normalized by subject weight
Rating HGS everyday Rating HGS activities Rating CSHT	Subjective ratings (see description in Section 2.2)

Table 1. *Cont.*

Variable	Description
Forearm circumference ndH Forearm circumference dH Upper arm circumference ndH Upper arm circumference dH	Forearm and upper arm perimeters
Length 2 ndH Length 3 ndH Length 4 ndH Length 5 ndH Length 2 dH Length 3 dH Length 4 dH Length 5 dH	Individual finger length: 2 represents the index finger, 5 the little finger
n pull-ups	Maximal number of pull-ups
CSHT	Climbing-specific holding time, measured performing dead hang on finger board

The evaluation process was integrated into a leave-one-out cross-validation procedure for the evaluation of the model. The features were individually scaled through removal of the mean and scaling to unit variance based on each training set. Calculations were performed in Python (Python Software Foundation, Wilmington, DE, USA) and SPSS Statistics (version 25, SPSS Inc., Chicago, IL, USA).

3. Results

Subjects achieved slightly higher HGS scores for the dominant hand (41.69 ± 12.15 kg) compared with the non-dominant hand (38.99 ± 12.12 kg). On average, subjects rated their subjective estimated CSHT as 4.40 ± 1.73 on an 11-point Likert scale. Perceived exhaustion after treatment was rated moderately by subjects, at 5.06 ± 2.15 on the OMNI scale. The correlations of all measured variables are presented in Figure 2.

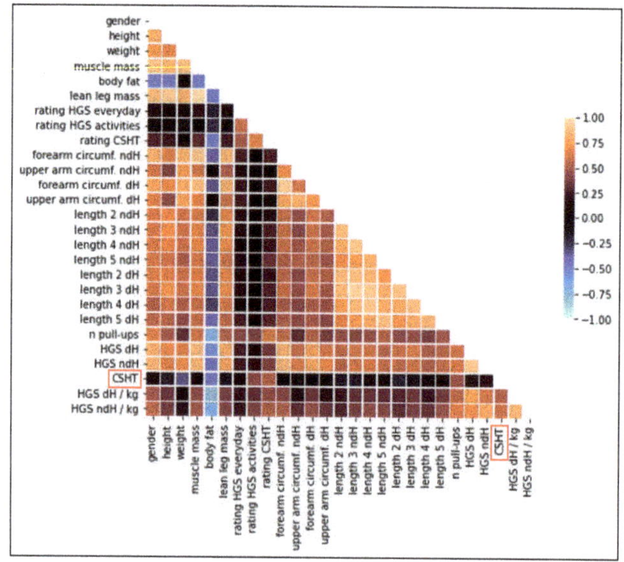

Figure 2. Descriptive correlation heat map of the measured variables. The color saturation represents the correlation strength (blue = negative correlation, yellow = positive correlation). The maximal climbing-specific holding time (CSHT) is highlighted with red boxes.

Using backward elimination separately for the dH and the ndH models, the same predictors were selected (weight, number of pull-ups, HGS, and length of middle finger). Multiple linear regression results showed that predictors demonstrated an effect on the climbing-specific holding time (dH: $F_{(4,43)} = 15.41$, $p < 0.001$, $n = 48$, $R^2 = 0.59$, $R^2_{corr} = 0.55$; ndH: $F_{(4,43)} = 21.38$, $p < 0.001$, $n = 48$, $R^2 = 0.67$, $R^2_{corr} = 0.63$). Detailed information regarding the predictors is presented in Table 2.

Table 2. Regressors of the multivariate model to predict the maximal climbing-specific holding time (CSHT) for dominant (dH) and non-dominant (ndH) hands. In addition, the regression coefficients, B; standardized coefficients, β; standard error, SER (quality of the estimate); and *p*-values of the respective predictors are shown. HGS = hand grip strength.

Predictors	ndH $R^2_{corr} = 0.63$				dH $R^2_{corr} = 0.55$			
	B	β	SER	*p*	B	β	SER	*p*
Constant	60.94	-	18.82	0.002	66.84	-	21.45	0.003
Weight	−0.47	−0.35	0.15	0.003	−0.44	−0.33	0.17	0.013
Number of pull-ups	1.17	0.35	0.41	0.006	1.63	0.49	0.43	0.000
HGS/kg	33.66	0.55	7.31	0.000	22.74	0.39	7.33	0.003
Length 3 (middle finger length)	−7.38	−0.31	2.83	0.013	−7.15	−0.30	3.27	0.034

Using machine learning for predictive modeling, an R^2 score of 0.51 was achieved during cross-validation. Comparing true vs. predicted values (Figure 3a), we found that the model underestimated predicted values for subjects with a relatively high CSHT. The features in relation to the coefficients of the model are presented in Figure 3b. Most of the most important features regarding their coefficients matched with the variables in the final regression models. For example, sex was unanimously assigned no relevance by excluding it or receiving a score close to zero. The same applied, for example, to the parameter HGS ndH/kg, which showed a height effect for both approaches.

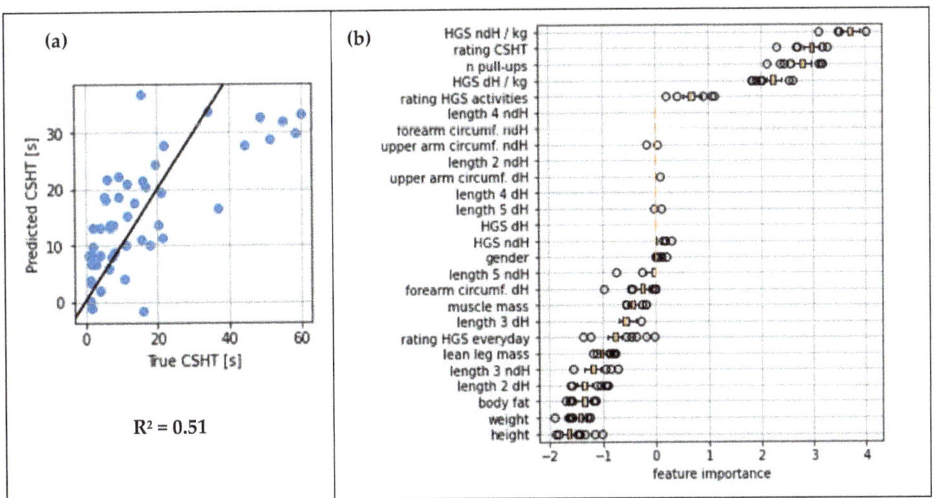

Figure 3. Results of the elastic net model. (**a**) True vs. predicted values for the CSHTS, as well as fit line for perfect fit. (**b**) Boxplots sorted according to the median coefficients (feature importance scores) of the features in the elastic net model during LOGO. Negative values represent an effect of high feature values reducing the CSHT; positive values represent an effect of high feature values increasing the CSHT. Please refer to Table 2 for feature labels.

As we observed the importance of the subjective rating of the CSHT in Figure 3b, the additional inclusion of this parameter was evaluated in the regression analysis, as this aspect was previously not included. The fit slightly improved when including the subjective rating of CSHT (dH: $F(5,42) = 14.44$, $p < 0.001$, $n = 48$, $R^2 = 0.63$, $R^2_{corr} = 0.59$; ndH: $F(5,42) = 19.32$, $p < 0.001$, $n = 48$, $R^2 = 0.70$, $R^2_{corr} = 0.66$).

4. Discussion

The results of the current study showed that, of the predictors evaluated, weight, maximal number of pull-ups, HGS normalized by subject weight, and length of the middle finger had an especially significant effect on the maximal CSHT. The high agreement of the regression analysis between the dominant and non-dominant sides could be positively interpreted regarding the quality of the results. We found a slightly better model fit for the analysis of the ndH, which may indicate that aspects of the ndH are more limiting regarding the CSHT compared with aspects of the dH. This is probably related to the expected finding that subjects achieved higher HGS scores for the dominant hand compared with the non-dominant hand. It should be noted that a higher symmetry of HGS was found in climbers compared with non-climbing subjects [22].

The results also showed that HGS alone seems to be of little use in predicting CSHT, although some studies in the field of climbing refer to this parameter (e.g., [23]). Instead, HGS should be normalized to individual body weight, as shown by the high relevance of this parameter in both approaches. However, despite the inclusion of various other possibly relevant factors, as well as complex, non-linear modeling techniques derived from the field of machine learning, we could only obtain an R^2_{corr} value of 0.63 for the non-dominant hand and an R^2_{corr} of 0.55 for the dominant hand for the regression analysis and an R score of 0.51 for the predictive modeling using the elastic net model. According to [12], a high relative grip strength is essential for climbing routes of high difficulty, but high relative grip strength does not necessarily relate to higher climbing performance. Furthermore, comparing EMG measurements, the authors in [24] showed that HGS measurements using hand dynamometers showed a lack of specificity in mapping sports climbing characteristics. The authors in [3] explained the weak association between HGS and climbing-specific performance with the lack of specificity in most hand positions that are required during sports climbing (except pinch grip). Further research should therefore aim to apply feature engineering and develop optimal experimental designs to identify better predictive variables to improve model accuracy. Furthermore, the authors in [25] reported that grip strength depended on many more variables, such as bone density, VO_{2max}, and the android-gynoid-fat-ratio, which we were not able to measure in this study, but which may also be related to CSHT. Additionally, Ref. [26] reported that climbing performance depends on flexibility, especially in the shoulder area. These variables were not measured in the current study and may be useful for improving model accuracy.

Finger length appeared to be a relevant factor for CSHT in the regression model results. As finger length is genetically determined, this could also be a crucial sub-factor when talent scouting. Therefore, the contributions of individual finger strength must be considered. The current findings, which showed that the length of the middle finger was a relevant parameter, are in line with a previous study [27] that explained that the middle finger produced the greatest force in hanging experiments. On the contrary, Ref. [28] found that ring finger strength was the most important individual finger strength measure for predicting climbing performance. In sports practice, it is sometimes suggested that the length of the little finger is crucial, as it affects how effectively the little finger can be used to hold crimps. In line with this notion, Ref. [27] reported that the little finger takes little more than $1/10$ of the absolute force on climbing holds. Missing $1/10$ of force can be significant in climbing, especially when climbing to exhaustion. However, the results of this study do not support the importance of the length of the little finger, as this parameter was not important in the final regression model. In the current study, some subjects were not able to keep their little finger on the holds of the fingerboard for a few seconds before termination, which could be explained by their progressive exhaustion during the dead hangs, leading to an inability to maintain the required hand position [3].

By identifying and adding the subjective rating of CSHT as a variable, the fit of the regression model could be further improved. Subjects seemed to be fairly good at assessing their abilities regarding the target parameter. Overall, the application of the ML model has shown great promise in expanding a background-knowledge-based regression model with data-driven knowledge concerning additional relevant parameters. Furthermore, we could confirm the suitability of the selected variables in the final regression model by the high level of agreement between the two approaches regarding the relevant parameters. This promising combination has also been demonstrated by other studies. In the context of cycling, for example, a combination of inferential statistical regression analysis and elastic net model has been used [29].

Both the regression model and the ML approach agree that sex had no relevant effect on the CSHT. This was surprising as sex differences were reported in HGS research [6,11]. Nevertheless, the results showed good fit with studies analyzing the forearm muscles during intermittent handgrip contractions, which is more specific to actual climbing loads, where they found no sex-specific differences in fatigability [30].

The authors in [26] showed that climbing performance was mainly related to training components, followed by anthropometric and flexibility components. Therefore, training status determines the CSHT to a significant degree. However, as found within the framework of the regression model, a person can also achieve a long holding time simply because of their low body weight. Accordingly, it cannot be generally said that people with longer CSHT are better-trained. To improve CSHT, the results suggest reducing body weight, as well as improving HGS and pull-up performance. However, climbing is characterized by a wide spectrum of conditional and coordinative abilities that are relevant to performance [31]. Therefore, isolated training of the parameters found appears to be insufficient for an optimal training strategy. According to [31], the best effect when undertaking conditioning training in climbing is obtained through a mixture of dynamic and static exercises in a semi-specific setting, combining hypertrophy, maximum strength, and endurance. In terms of more climbing-specific training, Ref. [32] states that interval bouldering works best as a form of sport-specific conditioning training, which may, in parallel, show positive effects on HGS and pull-up performance. In parallel with this approach, as shown by the results, attention should be paid to the athlete's weight.

An underestimation of the predicted value for subjects with a relatively high CSHT was found for the elastic net model. This may be related to an underrepresentation of these subjects regarding their CSHT values in the training data. An extension of the training data through the further inclusion of related subjects could thus contribute to improving the model.

In the current study, we used CSHT as an indicator of climbing performance, as strong associations have been reported in the literature [4]. However, it is important to keep in mind that we can only make indirect statements about actual performance with this procedure. It is therefore unclear whether the analyzed factors predict actual climbing performance better than CSHT. Additionally, CSHT does not necessarily represent a realistic climbing situation, as the sport involves grip changes and short pauses between isometric contractions. Furthermore, the legs support a large amount of the climbers' weight in a realistic climbing situation. In addition to the impact of anthropometric and fitness constitution, training age could also influence climbing performance. Research is therefore necessary to check if our results can be applied to more realistic climbing situations. In addition, other methods for the assessment of climbing-specific finger flexor strength, e.g., using a scale platform to measure the load that can be held by the test arm [33], may be helpful in gaining new insights.

5. Conclusions

The results of this study provide initial indications that finger length may play a role among many factors that map climbing talent in the context of talent scouting, but should not be seen in isolation from other potentially significant factors, such as technique or cognitive components. Furthermore, the results also confirm the performance-related importance of body weight, and thus, correspond to a general trend in competitive climbing, with extremely low body weights having been observed among these athletes and having been critically discussed in the context of anorexia athletica [34]. In the context of sports climbing training practice, the results indicate the potential relevance of regularly measuring the parameters of weight, maximum number of pull-ups, HGS normalized to body weight, and subjective assessment of CSHT to quantify training changes. Overall, the results support the potential usefulness of machine learning models as an exploratory tool in the context of sports science issues for enhancing inferential statistical analyses regarding the evaluation of predictive power, as well as the exploration and extension of existing knowledge.

Author Contributions: Conceptualization, C.D., E.B., S.B. and J.D.; methodology, C.D.; software, C.D.; validation, C.D., E.B. and J.D; formal analysis, C.D., E.B., O.L. and J.D.; investigation, J.D. and A.M.; resources, M.F.; data curation, C.D.; writing—original draft preparation, C.D., E.B., O.L., M.S., and J.D.; writing—review and editing, C.D., E.B., O.L., M.S., S.B., M.F. and J.D.; visualization, C.D.; supervision, O.L. and M.F.; project administration, M.F.; funding acquisition, M.F. All authors have read and agreed to the published version of the manuscript.

Funding: This research was funded by Offene Digitalisierungsallianz Pfalz, BMBF [03IHS075B].

Institutional Review Board Statement: The study was conducted in accordance with the Declaration of Helsinki and approved by the Ethics Committee of Technische Universität Kaiserslautern (protocol code: 43; date of approval: 25.11.2021).

Informed Consent Statement: Informed consent was obtained from all subjects involved in the study. Written informed consent has been obtained from the participants to publish this paper.

Data Availability Statement: The data are available if there is justified research interest.

Acknowledgments: Foremost, the authors would like to thank all of the participants of this study.

Conflicts of Interest: The authors declare no conflict of interest. The funders had no role in the design of the study; in the collection, analyses, or interpretation of data; in the writing of the manuscript; or in the decision to publish the results.

References

1. MacKenzie, R.; Monaghan, L.; Masson, R.A.; Werner, A.K.; Caprez, T.S.; Johnston, L.; Kemi, O.J. Physical and Physiological Determinants of Rock Climbing. *Int. J. Sports Physiol. Perform.* **2020**, *15*, 168–179. [CrossRef] [PubMed]
2. Salehhodin, S.N.; Abdullah, B.; Yusoff, A. Comparison Level of Handgrip Strength for the Three Categories among Male Athleteâs Artificial Wall Climbing and Factors WILL Affect. *Int. J. Acad. Res. Bus. Soc. Sci.* **2017**, *7*, 272–285. [CrossRef]
3. Watts, P.B. Physiology of difficult rock climbing. *Eur. J. Appl. Physiol.* **2004**, *91*, 361–372. [CrossRef] [PubMed]
4. Bergua, P.; Montero-Marin, J.; Gomez-Bruton, A.; Casajús, J.A. Hanging ability in climbing: An approach by finger hangs on adjusted depth edges in advanced and elite sport climbers. *Int. J. Perform. Anal. Sport* **2018**, *18*, 437–450. [CrossRef]
5. Jandl, D. Auswirkungen von Zwei Griffbrett-Trainingsprotokollen Auf Die Kraftentwicklung der Fingerbeugermuskulatur Beim Klettern. Diploma Thesis, Karl-Franzens-Universität Graz, Graz, Austria, 2021.
6. Hahn, P.; Spies, C.; Unglaub, F.; Mühldorfer-Fodor, M. Die Messung der Griffkraft: Wertigkeit und Grenzen. *Orthopade* **2018**, *47*, 191–197. [CrossRef]
7. Watts, P.B.; Martin, D.T.; Durtschi, S. Anthropometric profiles of elite male and female competitive sport rock climbers. *J. Sports Sci.* **1993**, *11*, 113–117. [CrossRef]
8. Booth, J.; Marino, F.; Hill, C.; Gwinn, T. Energy cost of sport rock climbing in elite performers. *Br. J. Sports Med.* **1999**, *33*, 14–18. [CrossRef]
9. Watts, P.B.; Daggett, M.; Gallagher, P.; Wilkins, B. Metabolic response during sport rock climbing and the effects of active versus passive recovery. *Int. J. Sports Med.* **2000**, *21*, 185–190. [CrossRef]
10. Chau, N.; Pétry, D.; Bourgkard, E.; Huguenin, P.; Remy, E.; André, J.M. Comparison between estimates of hand volume and hand strengths with sex and age with and without anthropometric data in healthy working people. *Eur. J. Epidemiol.* **1997**, *13*, 309–316. [CrossRef]

11. Ekşioğlu, M. Normative static grip strength of population of Turkey, effects of various factors and a comparison with international norms. *Appl. Ergon.* **2016**, *52*, 8–17. [CrossRef]
12. Baláš, J.; Pecha, O.; Martin, A.J.; Cochrane, D. Hand–arm strength and endurance as predictors of climbing performance. *Eur. J. Sport Sci.* **2012**, *12*, 16–25. [CrossRef]
13. Teufl, W.; Taetz, B.; Miezal, M.; Dindorf, C.; Fröhlich, M.; Trinler, U.; Hogan, A.; Bleser, G. Automated detection and explainability of pathological gait patterns using a one-class support vector machine trained on inertial measurement unit based gait data. *Clin. Biomech.* **2021**, *89*, 105452. [CrossRef] [PubMed]
14. Horst, F.; Lapuschkin, S.; Samek, W.; Müller, K.-R.; Schöllhorn, W.I. Explaining the unique nature of individual gait patterns with deep learning. *Sci. Rep.* **2019**, *9*, 2391. [CrossRef] [PubMed]
15. Duh, Y.-S.; Chang, R. Recurrent Neural Network for MoonBoard Climbing Route Classification and Generation. *arXiv* **2021**, arXiv:2102.01788.
16. Ambrasat, J.; Schupp, J. *Handgreifkraftmessung im Sozio-Oekonomischen Panel (SOEP) 2006 und 2008*; Deutsches Institut für Wirtschaftsforschung (DIW Berlin): Berlin, Germany, 2011.
17. Mathiowetz, V.; Weber, K.; Volland, G.; Kashman, N. Reliability and validity of grip and pinch strength evaluations. *J. Hand Surg.* **1984**, *9*, 222–226. [CrossRef]
18. Amca, A.M.; Vigouroux, L.; Aritan, S.; Berton, E. The effect of chalk on the finger-hold friction coefficient in rock climbing. *Sports Biomech.* **2012**, *11*, 473–479. [CrossRef]
19. Robertson, R.J. *Perceived Exertion for Practitioners: Rating Effort with the OMNI Picture System*; Human Kinetics: Leeds, UK, 2004; ISBN 9780736048378.
20. Pedregosa, F.; Varoquaux, G.; Gramfort, A.; Michel, V.; Thirion, B.; Grisel, O.; Blondel, M.; Prettenhofer, P.; Weiss, R.; Dubourg, V.; et al. Scikit-learn: Machine learning in Python. *J. Mach. Learn. Res.* **2011**, *12*, 2825–2830.
21. Zou, H.; Hastie, T. Regularization and variable selection via the elastic net. *J. R. Stat. Soc. Ser. B* **2005**, *67*, 301–320. [CrossRef]
22. Grant, S.; Hynes, V.; Whittaker, A.; Aitchison, T. Anthropometric, strength, endurance and flexibility characteristics of elite and recreational climbers. *J. Sports Sci.* **1996**, *14*, 301–309. [CrossRef]
23. Gürer, B.; Yıldız, M.E. Investigation of Sport Rock Climbers' Handgrip Strength. *J. Biol. Exerc.* **2015**, *11*, 55–71. [CrossRef]
24. Watts, P.; Jensen, R.; Gannon, E.; Kobeinia, R.; Maynard, J.; Sansom, J. Forearm EMG During Rock Climbing Differs from EMG During Handgrip Dynamometry. *Int. J. Exerc. Sci.* **2008**, *1*, 4–13.
25. Pratt, J.; De Vito, G.; Narici, M.; Segurado, R.; Dolan, J.; Conroy, J.; Boreham, C. Grip strength performance from 9431 participants of the GenoFit study: Normative data and associated factors. *GeroScience* **2021**, *43*, 2533–2546. [CrossRef] [PubMed]
26. Mermier, C.M.; Janot, J.M.; Parker, D.L.; Swan, J.G. Physiological and anthropometric determinants of sport climbing performance. *Br. J. Sports Med.* **2000**, *34*, 359–365. [CrossRef]
27. Herman-Dunphy, D.; King, D.L. The Effects of Modern Climbing Holds on the Finger Forces. *Int. J. Exerc. Sci.* **2017**, *9*, 46.
28. Stefan, R.R.; Camic, C.L.; Miles, G.F.; Kovacs, A.J.; Jagim, A.R.; Hill, C.M. Relative Contributions of Handgrip and Individual Finger Strength on Climbing Performance in a Bouldering Competition. *Int. J. Sports Physiol. Perform.* **2022**, *17*, 768–773. [CrossRef]
29. Molenaar, L. Performance of General Classification Riders on a Final Climb in Professional Road Cycling: What Can Be Predicted and What Does Influence Performance? Diploma Thesis, University of Groningen, Groningen, The Netherlands, 2020.
30. Gonzales, J.U.; Scheuermann, B.W. Absence of gender differences in the fatigability of the forearm muscles during intermittent isometric handgrip exercise. *J. Sports Sci. Med.* **2007**, *6*, 98–105.
31. Langer, K.; Simon, C.; Wiemeyer, J. Strength Training in Climbing: A Systematic Review. *J. Strength Cond. Res.* **2022**, *Publish Ahead of Print*. [CrossRef]
32. Medernach, J.P.; Kleinöder, H.; Lötzerich, H.H.H. Effect of interval bouldering on hanging and climbing time to exhaustion. *Sports Technol.* **2015**, *8*, 76–82. [CrossRef]
33. Baláš, J.; Mrskoč, J.; Panáčková, M.; Draper, N. Sport-specific finger flexor strength assessment using electronic scales in sport climbers. *Sports Technol.* **2014**, *7*, 151–158. [CrossRef]
34. Lutter, C.; El-Sheikh, Y.; Schöffl, I.; Schöffl, V. Sport climbing: Medical considerations for this new Olympic discipline. *Br. J. Sports Med.* **2017**, *51*, 2–3. [CrossRef]

Journal of
Functional Morphology and Kinesiology

Article

Evaluation of Posturographic and Neuromuscular Parameters during Upright Stance and Hand Standing: A Pilot Study

Ewan Thomas [1], Carlo Rossi [1,*], Luca Petrigna [2], Giuseppe Messina [1], Marianna Bellafiore [1], Fatma Neşe Şahin [3], Patrizia Proia [1], Antonio Palma [1] and Antonino Bianco [1]

[1] Sport and Exercise Sciences Research Unit, Department of Psychology, Educational Science and Human Movement, University of Palermo, Via Giovanni Pascoli 6, 90144 Palermo, Italy; ewan.thomas@unipa.it (E.T.); giuseppe.messina17@unipa.it (G.M.); marianna.bellafiore@unipa.it (M.B.); patrizia.proia@unipa.it (P.P.); antonio.palma@unipa.it (A.P.); antonino.bianco@unipa.it (A.B.)

[2] Department of Biomedical and Biotechnological Sciences, Section of Anatomy, Histology and Movement Science, School of Medicine, University of Catania, Via S. Sofia n°97, 95123 Catania, Italy; luca.petrigna@unict.it

[3] Department of Coaching Education, Faculty of Sport Science, Ankara University, Ankara 06830, Türkiye; nese.sahin@ankara.edu.tr

* Correspondence: carlo.rossi@unipa.it; Tel.: +39-388-6311467

Abstract: Upright bipedal posture is the physiological human posture; however, it is not the only possible form of human standing; indeed, an inverted position, a handstand, is required during gymnastics or other sports. Thus, this study aimed to understand the differences between the two standing strategies from a postural and neuromuscular perspective. Thirteen gymnasts with at least three years of sports experience underwent a baropodometric assessment and a surface electromyography (sEMG) examination in a standard upright bipodalic stance and during a handstand. The sEMG examination was performed on the gastrocnemius during an upright stance and on the flexor carpi radialis during the handstand. Limb weight distribution presented differences between the two vertical stances ($p < 0.01$). During the handstand, the weight ratio was prevalently observed on the palm of the hand for both hands with a significant difference between the front and rear aspect of the hand compared to the standing tasks ($p < 0.01$). Normalized sEMG amplitude showed significant differences during bipedal standing and hand standing; however, over a 5 s period, the normalized median frequency (MDF) value was similar for the two tasks. Both standing tasks presented similar postural weight managing patterns when analysed on the frontal plane, but they were different on the sagittal plane. In addition, the neuromuscular patterns during a 5 s window differ in amplitude but not for the frequency domain.

Keywords: handstand; postural control; postural balance; sEMG; stabilometric assessment; exercise

Citation: Thomas, E.; Rossi, C.; Petrigna, L.; Messina, G.; Bellafiore, M.; Şahin, F.N.; Proia, P.; Palma, A.; Bianco, A. Evaluation of Posturographic and Neuromuscular Parameters during Upright Stance and Hand Standing: A Pilot Study. *J. Funct. Morphol. Kinesiol.* **2023**, *8*, 40. https://doi.org/10.3390/jfmk8020040

Academic Editor: Olivier Hue

Received: 6 February 2023
Revised: 24 March 2023
Accepted: 28 March 2023
Published: 30 March 2023

1. Introduction

Postural control is the ability of the nervous system to modulate the distribution and geometry of mass, as well as position and orientation through the integration of information from the vestibular, visual, and tactile apparatus [1]. Humans develop balance control strategies in postural positions at an early age by arranging muscle contractions to create torques around joints to preserve the center of mass (CoM) projection within the region of stability at the base of the support [2,3]. However, postural perturbation experiments have shown that ankle, hip, ankle-hip, and ankle-knee-hip strategies can be rapidly organized into a reactive mode to preserve balance stability [4,5]. It must be considered that bipedal postural control is not the only posture; in fact, gymnasts and people who work in the circus also adopt other forms of standing positions, such as hand standing [6]. As with the bipedal posture, the hand standing also integrates information from vision, vestibules, and tactile resolution of the hands [1,7]. The handstand is a fundamental skill in gymnastics in both men's and women's and is also included in more complex gymnastic skills [8]. The limited stability region of the handstand leads to a requirement for a balance control strategy that is more constrained than that for standing [1,9,10]. In addition, the handstand motion is a less natural motion, which can cause a feeling of unsteadiness or disorientation which results in the need for a more constrained balance control strategy than standing [4]. Thus, the main difference between the two types of postures could be due to the different muscles and joints adopted and, consequently, to the different strategies that the body has to adopt to maintain such posture. The hand standing posture, like the bipedal standing posture, also adopts three joint levels [1]. During handstands, it has been noted that the body begins to move earlier in the distal body segments while the trunk and head remain more stable [10]. Similarly, during the bipedal position, the ankle is the first joint which gets involved, while, during the handstand, the activation of the wrist flexors is the primary strategy to maintain balance [11]. This could be explained by the closer position of the wrist to the contact point on the ground and the need to minimize upper body movement [12]. Apparently, the best balances during the handstand are represented by major contributions from wrist and shoulder torques with little influence from hip torques. On the contrary, there is increasing awareness regarding the importance of hip torques, which are more influential in less successful balances [7,12]. In regards to the bipedal posture [13], the experience carried out with gymnastics athletes, young people, and adults have demonstrated a different relationship between muscle activity and postural control variables; furthermore, there is a reduction of the pressure shift and, consequently, less muscle activity related to the use of the wrist flexors together with the anterior deltoid [11].

The two tasks share common processes such as the sense and movement of the body and the implementation of a multi-joint strategies to preserve the orientation within certain constraints. Consequently, understanding mechanisms or strategies in one task can lead to insights into the other [7]. Additionally, people use a small set of preferred compensatory movement strategies to preserve balance during the handstand [10].

One study using two support surfaces, one foam and the other solid, highlighted the difficulty that even experienced people have in controlling performance during hand standing on non-solid surfaces [14]. Furthermore, signal delays are an important feature of any biological system, with large delays reducing stability and complicating control, and, consequently, this will be evaluated through surface electromyography (sEMG). Handstand balance performed by experienced gymnasts provides an alternative perspective to a normal standing position for understanding this complex system. Therefore, the present study aimed to compare the bipedal position and the hand standing position by evaluating the two tasks from a baropodometric and neuromuscular point of view in gymnasts. In this study, we wanted to understand the differences between the upright bipedal and vertical posture, in terms of body weight distribution and neuromuscular activation. We assumed that there would be differences in body weight distribution and neuromuscular patterns between the two postures.

2. Material and Methods

2.1. Participants

A total of 13 gymnasts with at least three years of sports experience, 10 males and 3 females (mean and standard deviation: age (years) 20.2 ± 4; height (cm) 170.4 ± 7.4; weight (kg) 66.0 ± 11.0), were retained for investigation. Participants were included if they were free of injuries during the assessment period and if they had an experience of at least three years in training with a handstand. Each gymnast regularly exercised ~1 h daily from Monday to Friday. Before the study, participants were informed about the study protocol, and the risks and benefits of their participation, and they gave written individual informed consent to participate in the study. The study was approved by the local Bioethics Committee of the University of Palermo (ref. n°121/2023) and was carried out according to the principles established by the Declaration of Helsinki.

2.2. Study Design

The total duration of each assessment lasted around 30 min, during which participants compiled and signed the documentation. Subsequently, anthropometric measurements were taken. Once all individuals' data were collected, each participant had to undergo a maximal voluntary contraction (MVC). Following the MVC, a sEMG evaluation was performed for the gastrocnemius and the flexor carpi radialis together with a baropodometric assessment. The documents signed by the participants were related to their consent to take part in the study and their permission to use their data. Anthropometric measures were related to height, evaluated with a meter, and weight, measured with a professional balance Seca scale (maximum weight recordable: 300 kg; resolution: 100 g; Seca, Hamburg, Germany). The baropodometric assessment was performed in an upright stance and during a handstand (Figure 1). For each task, the sEMG examination of the gastrocnemius during an upright stance and of the flexor carpi radialis during the handstand was retained for analysis. Before the assessment, participants performed a light warm-up consisting of 1 set of dynamic stretching of the targeted muscles of 30 s duration [15].

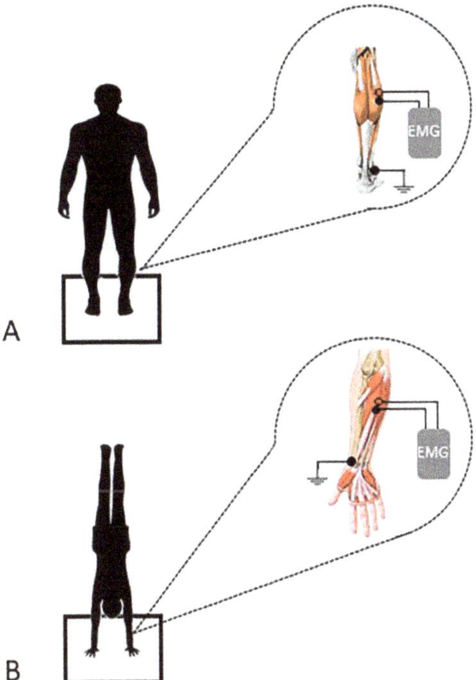

Figure 1. Graphical depiction of the experimental setting. In panel (**A**), the baropodometric evaluation and the sEMG applied to the gastrocnemius muscle during the standing task. In panel (**B**), the baropodometric evaluation and the sEMG applied to the flexor carpi radialis muscle during the hand standing task.

2.3. Baropodometric Assessment

All tasks were performed on a FreeMed baropodometric platform (50 × 60 cm) and with the FreeStep v.1.0.3 software. The sensors, coated with 24 K gold, guaranteed the repeatability and reliability of the instrument (produced by Sensor Medica, Guidonia Montecelio, Roma, Italy). During the upright stance, the gymnasts had to stand with feet parallel, and they had to keep that position for the duration of the assessment. The upper limbs were completely extended. Participants had to fixate on a point 2 m ahead, on a white surface, at eye level. The length of the trial was 5 s. No indications were given by the investigator related to the distance between the feet or the position to keep during the assessment not to influence the posture of each participant. However, reference points in the platform assessment area had to be maintained. Participants were barefoot.

During the handstand, the hand position was chosen by the participants. As above, reference points in the platform assessment area had to be maintained. The length of the trial was 5 s. Changes in support surface or the contact of the floor with the lower limbs was considered a failure of the task. If the participant failed the task, this could be repeated a maximum of 3 times after a 5 min break to avoid muscle fatigue.

Variables retained from the baropodometric assessment concerned: Support surface for each limb (cm^2) with information regarding the total support surface and the front and rear aspect of each analysed limb (based on the acquired image with a 50% ratio between front and rear aspect). Further, weight distribution (%) for the same variables mentioned above were also retained.

2.4. sEMG Evaluation

Before the sEMG evaluation, an MVC was performed. The maximal isometric force exerted by the gastrocnemius muscle and the flexor carpi radialis muscle were determined by asking the participants to increase the force from rest to maximum gradually in $\sim$3 s and to then maintain the maximum for an additional 3 s. Repeated contractions were performed until 2 attempts were within 5% of each other, and the greater peak force was used as the subject's MVC force [16]. A customized apparatus was built to perform the MVC of the gastrocnemius muscle and the flexor carpi radialis muscle.

Before the sEMG evaluation, participants' skin was shaved (if necessary) and cleaned with alcohol at 70%. The signals were collected using an OTBio Quattro EMG device (Copyright@ 2010 OT Bioelettronica, Torino, Italy), unilaterally, from the dominant side of the body using a bipolar method. The surface electrodes (Ag/AgCl sEMG disposable adhesive circular bipolar surface electrodes of 24 mm diameter, CDE02401500BX, Spes Medica S.r.L, Battipaglia, Italy), were placed with a distance of 1 cm between them on the bellies of the gastrocnemius and the flexor carpi radialis muscles. Electrode location was determined according to the recommendations provided by Barbero et al. [17] and the manufacturer's instructions.

The reference electrode (20 × 25 mm snap disposable electrodes, DENIS, Spes Medica S.r.L, Battipaglia, Italy) was placed on the malleolus of the fibula during the evaluation of the gastrocnemius muscle and on the styloid process of the radius during the evaluation of the flexor carpi radialis muscle (Figure 1).

All sEMG data was bandpass filtered using cut-off frequencies between 20 and 400 Hz. The signal was then processed to obtain the amplitudes root mean square (RMS) and the median frequency (MDF) for either the standing or hand-standing task. The obtained values were then normalized according to each participant's MVC for each task.

2.5. Statistical Analysis

Means, standard deviations, and 95% confidence intervals (CI) were calculated to present data. Inferential statistics were carried out with Jamovi (The jamovi project (2021). jamovi (Version 1.8.0.1) [Computer Software]. Retrieved from https://www.jamovi.org). A Shapiro—Wilks test was performed to identify the normality of the distribution of all parameters. Parametric and non-parametric assessment was adopted when appropriate. To assess the differences in stabilometry and sEMG between tasks for each participant a paired *t*-test or a Wilcoxon test were adopted when appropriate. For each analysis, the effect size (ES) was calculated. Cohen δ's were adopted as measures of ES for parametric data, while biserial rank correlation was used for non-parametric data. The magnitude of the Cohen δ's ES was classified according to the following scale: 0–0.19 = trivial effect, 0.20–0.49 = small effect, 0.50–0.79 = moderate effect, and $\geq$0.80 = large effect [18], while the magnitude for the biserial rank correlation were classified as 0–0.1 = trivial effect, 0.10–0.3 = small effect,

0.30–0.5 = moderate effect, and ≥ 0.50 = large effect [19]. Graphs were created with Graph-Pad Prism8 (GraphPad Software, San Diego, CA, USA). The significance for all analyses was set at $p < 0.05$.

3. Results

Weight is equally distributed during the upright stance with no differences between the left and the right limb (50.3 ± 1.8% and 49.7 ± 1.8%, respectively). Similarly, also during hand standing, no statistically significant differences were detected in the weight distribution between left and right limbs (45.1 ± 4.9% and 54.9 ± 4.9%, respectively) (Figure 2). However, there were significant differences between the two tasks (upright stance vs. handstand) (Table 1).

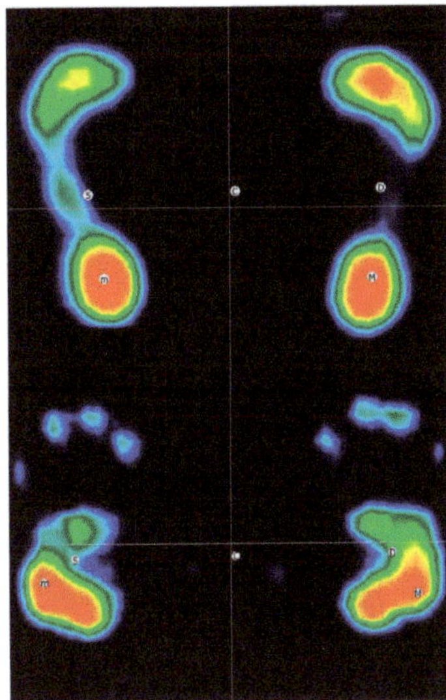

Figure 2. Representation of the standing and hand standing task during the baropodometric evaluation of one participant.

Table 1. Descriptive measures of the baropodometic assessment during standing and hand standing.

	Standing	Hand Standing	p	ES	CI (95%)-S	CI (95%)-HS
Left supporting surface (cm^2)	126.1 ± 22.7	86.6 ± 19.2	<0.01 §	1.75	112.5–139.8	75–98.2
Right supporting surface (cm^2)	122.1 ± 21.2	90.8 ± 15.2	<0.01 §	2.69	109.4–135	81.6–99.9
Left forefoot/hand (cm^2)	70.4 ± 13.8	32.9 ± 7.8	<0.01 §	2.70	62.1–78.7	28.2–37.7
Right forefoot/hand (cm^2)	67.8 ± 13.9	34.3 ± 8.1	<0.01 §	2.30	59.5–76.2	29.4–39.2
Left backfoot/hand (cm^2)	55.7 ± 10.1	53.7 ± 12.9	0.60 §	0.14	49.6–61.8	45.9–61.5
Right backfoot/hand (cm^2)	54.4 ± 9.3	56.4 ± 8.3	0.45 §	0.21	48.8–60	51.4–61.4
Left supporting surface (%)	50.3 ± 1.8	45.1 ± 4.9	<0.01 #	0.91	49.2–51.4	42.2–48.1
Right supporting surface (%)	49.7 ± 1.8	54.9 ± 4.9	<0.01 #	0.91	48.6–50.8	51.9–57.8
Left forefoot/hand (%)	49.1 ± 7.3	27.6 ± 6.1	<0.01 §	2.17	44.8–53.5	24.3–30.9
Right forefoot/hand (%)	47.8 ± 6.6	25.1 ± 5.4	<0.01 §	2.30	43.8–51.7	21.5–28.8
Left backfoot/hand (%)	50.8 ± 7.3	72.4 ± 6.1	<0.01 §	2.17	46.5–55.2	69.1–75.7
Right backfoot/hand (%)	52.2 ± 6.6	74.9 ± 5.4	<0.01 §	2.30	48.3–56.2	71.2–78.5

§-parametric evaluation; #-non-parametric evaluation; ES-effect size; CI-confidence interval; S-standing; HS-hand standing.

Regarding the analysis of the base of support during the upright stance, the forefoot and rear part of the foot were similar (F/R 49/51% weight distribution).

Conversely, during the hand-standing task, the ratio was prevalently on the palm of the hand (F/R 27/73% weight distribution). This difference was significant, and it was noted for both hands ($p < 0.01$).

Normalized sEMG amplitude (RMS) showed a significant difference ($p < 0.01$) during bipedal standing and hand standing (5.9 $\pm$ 2.6 vs. 92.9 $\pm$ 45.8% activation) (Figure 3). However, over a 5 s period, the normalized MDF value (Figure 4) was similar for the two tasks (71.2 $\pm$ 18.5 vs. 71.9 $\pm$ 14.0, standing vs. hand standing, respectively) (Table 2).

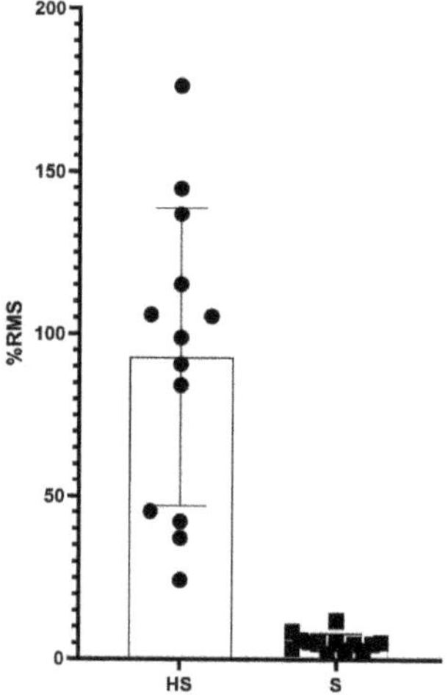

Figure 3. Figure represents the mean and studv of sEMG amplitudes RMS (root mean square) percentages of the two performed tasks of the analyzed sample HS = Handstand; S = Standing.

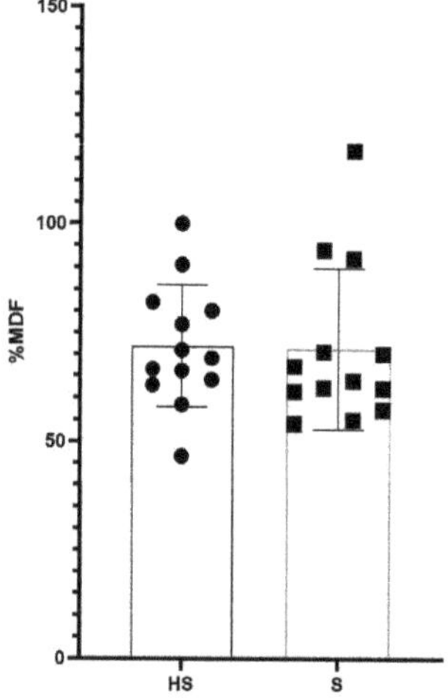

Figure 4. Figure represents the mean and study of sEMG MDF (median frequency) percentages of the two performed tasks of the analyzed sample. HS = Handstand; S = Standing.

Table 2. sEMG Measures of both standing and hand standing.

	Standing	Hand Standing	p	ES	CI (95%)-S	CI (95%)-HS
MVC (µV)	194.6 ± 147.9	405.6 ± 200.6	0.01 #	0.78	105–284	284–527
Mean Amplitude (µV)	11.5 ± 12.6	326.1 ± 145.3	<0.01 #	1.00	3.9–19.2	238–414
MDF (Hz)	164.4 ± 30.0	99.3 ± 28.3	<0.01 §	1.54	146–183	82.2–116
Mean Frequency (Hz)	114.4 ± 23.2	69.1 ± 15.1	<0.01 §	2.13	100.3–128.4	60–78
Normalized Amplitude (%RMS)	5.2 ± 2.6	92.9 ± 45.8	<0.01 §	1.90	3.6–6.8	65.2–120.6
Normalized Frequency (%MDF)	71.2 ± 18.5	71.9 ± 14.0	0.54 #	0.21	60–82.4	63.4–80.3

§-parametric evaluation; #-non-parametric evaluation; ES-effect size; CI-confidence interval; S-standing; HS-hand standing.

4. Discussion

The purpose of the following study was to understand the differences between the standing strategy and the hand standing from a postural and neuromuscular perspective and to evaluate if any differences were present. The main results of this study highlight similarities between the two tasks regarding weight distributions between the two limbs, with differences within each limb across tasks. Concerning the normalized sEMG amplitude, there were significant differences during the bipedal position and the hand standing. The handstand is considered a fundamental gymnastic skill [8]. On the other hand, the variability of acceleration during the inverted positions could serve as a source of information to preserve stability [10]. For weight management strategies in the sagittal plane, we found a difference between the two tasks. Furthermore, the neuromuscular patterns during a 5-s window differ in the time-intensity domain but not in the frequency domain. The ability to orient body parts concerning gravity, support surface, visual environment, and internal references is a critical component of the postural control [2]. A healthy nervous system automatically changes the way the body is oriented in space, depending on the context and the task [2].

The level of experience of the sportsman considerably differentiates the stability of an upright body. In fact, subjects with greater experience who practice gymnastics, compared to young people, are characterized by a better ability to control the position of the body in both positions [20]. Hand standing has been seen to provide a larger center of pressure (CoP) area than the upright postural position, which could be caused by a smaller contact area which reduces the base of support of the body, and a reduction in postural control due to the unusual balancing task [10,21]. In addition, hand stand experience has been associated with better performances of postural tasks [14,22]. Based on the literature on bipedal initiation, postural adjustments that help achieve optimal balance and propulsive requirements for gait execution are performed in the movement preparation phase [23–27]. Grabowiecki et al., 2021 proposed that the anterior displacement strategy is implemented when the CoP is closest to the posterior boundary of the BoS (base of support) and behind the vertical projection of the CoM at the onset of walking initiation during handstand [28]. Conversely, the posterior displacement strategy is performed when the CoP position is beyond the vertical projection of the CoM and closer to the anterior border of the BoS.

As a result, the reduced surface area of the hand acts like the foot to maintain body balance with the wrists and shoulders acting as the ankle and hip (strategy adopted during bipedal stance) [11]. It is well-known that keeping a straight body shape without any angles in the shoulder, elbow, hip, and knee joints and a strong balance between agonists and antagonist's muscles is required for high-quality handstand postures [29–31]. Therefore, such findings highlight that shoulder torque plays a more important role than hip torque in the handstand postural control strategy [7,12,32]. Concerning "ankle strategy" and "hip strategy" in standing [33], the hip specificities of gymnasts seem relatively uncoordinated and arbitrary at least until the wrists and shoulders work mainly to regulate postural balance [22]. Therefore, future research on this topic should consider a study on the "wrist strategy".

We also found significant values for the forefoot and hindfoot, both right and left with the respective hand. In the study by Yamazaki et al., 2005 it was seen that the trajectories of the foot pressure center varied, initially towards the Rt side and then towards Lt which, respectively, coincided with the initial and subsequent phases of the trunk rotations and muscle activation [34]. Postural adjustment in the handstand would appear to be organized according to a system similar to that of upright posture, with joint levels suggesting the existence of an organization typical of human posture [35]. The results of this study can provide important insights into human motor performance in different positions and how postural and neuromuscular factors interact with each other. This can be especially useful for athletes and professionals who require exceptional postural stability and superior neuromuscular strength.

Regarding the sEMG amplitude, we observed a significant difference between the standing and hand standing tasks. It was expected that such differences were present, since hand standing is a complex skill. The results obtained by the present investigation concerning the high activation of the wrist flexors are similar to those of Kochanowicz et al. [1]. The wrist flexors in the context of hand standing act like the plantar flexors during conventional standing [1]. On the other hand, the sEMG amplitude during quiet standing showed very low values, which are in line with existing literature for resting muscle activity during balancing tasks [36]. In the frequency domain, we found no significant differences between the hand standing and the standing tasks, unlike the study by Wyatt et al. 2021, which, instead, showed through a multiple regression that the frequency domain is an excellent predictor of the duration of the handstand equilibrium [37]. This is a metric that is usually used to identify patterns of fatigue. Probably, the five-second window adopted in the present study is too short to cause fatigue [38]. Secondly, the sample was composed of expert gymnasts which are used to performing the handstand task, therefore preventing any variation in the frequency discharge of the involved muscles [11,20]. Recruitment strategies at the time of measurement were similar for upper and lower limbs during static postural stability tasks [39].

Their results showed that the alternating rotations of the upper torso, produced by rapid arm movements, were transmitted to the hip in part due to co-contraction of the trunk muscles, and each pair of muscles in the hip joint contributed to maintaining upright posture by stabilizing the hip joints against alternating rotations of the trunk [35]. The percentage of time spent in different control strategies for perturbed and unperturbed standing and handstand balance was determined [4], during both perturbed and unperturbed balance, the predominant control strategies were a standing ankle strategy and a vertical wrist strategy. Findings reveal that the central nervous system maintains balance during a variety of tasks and postures by employing an individual control strategy [4].

Limitations of the present study are the small number of recruited participants and also the specific population analyzed concerning only young gymnasts. Another limit of the study was that no control was performed on athletes' previous injuries. Future studies should also include a greater proportion of women, other age groups, and people with different sporting backgrounds who can perform a handstand.

5. Conclusions

Postural and neuromuscular differences were observed between standing and hand standing. During a handstand, we observed that the majority of the support surface was attributed to the area of the palm of the hand, while an even distribution between hind foot and rear foot was observed during the standing task. Furthermore, sEMG muscle activity was different between the two positions with greater effort during hand standing vs. standing. However, analysis of frequency was similar across the two positions over a 5 s time frame. More research is needed to gain insight into the different forms of standing postures.

Author Contributions: Conceptualization, E.T. and L.P.; methodology, E.T. and M.B.; formal analysis, M.B.; writing-original draft, E.T. and C.R.; review and editing, C.R. and P.P.; statistical analysis, F.N.Ş.; validation, C.R.; investigation, G.M.; resources, E.T.; data curation, C.R. and M.B.; writing, E.T.; visualization, E.T. and L.P.; supervision, A.P. and A.B.; project administration, A.B. All authors have read and agreed to the version of the manuscript.

Funding: This research received no external funding.

Institutional Review Board Statement: The study was conducted in accordance with the Declaration of Helsinki and approved by the Bioethics Committee of the University of Palermo (Protocol code 121/2023).

Informed Consent Statement: Informed consent was obtained from all subjects involved in the study.

Data Availability Statement: Data are available from the corresponding author upon reasonable request.

Conflicts of Interest: The authors declare no conflict of interest.

References

1. Gautier, G.; Thouvarecq, R.; Chollet, D. Visual and postural control of an arbitrary posture: The handstand. *J. Sports Sci.* **2007**, *25*, 1271–1278. [CrossRef]
2. Horak, F.B. Postural orientation and equilibrium: What do we need to know about neural control of balance to prevent falls? *Age Ageing* **2006**, *35*, ii7–ii11. [CrossRef]
3. Winter, D.A. Human balance and posture control during standing and walking. *Gait Posture* **1995**, *3*, 193–214. [CrossRef]
4. Blenkinsop, G.M.; Pain, M.T.; Hiley, M.J. Balance control strategies during perturbed and unperturbed balance in standing and handstand. *R. Soc. Open Sci.* **2017**, *4*, 161018. [CrossRef]
5. Creath, R.; Kiemel, T.; Horak, F.; Peterka, R.; Jeka, J. A unified view of quiet and perturbed stance: Simultaneous co-existing excitable modes. *Neurosci. Lett.* **2005**, *377*, 75–80. [CrossRef]
6. Vuillerme, N.; Danion, F.; Marin, L.; Boyadjian, A.; Prieur, J.M.; Weise, I.; Nougier, V. The effect of expertise in gymnastics on postural control. *Neurosci. Lett.* **2001**, *303*, 83–86. [CrossRef]
7. Yeadon, M.R.; Trewartha, G. Control strategy for a hand balance. *Mot. Control* **2003**, *7*, 421–442. [CrossRef]
8. Readhead, L. *Men's Gymnastics Coaching Manual*; Crowood Press: Ramsbury, UK, 1997.
9. Clément, G.; Rezette, D. Motor behavior underlying the control of an upside-down vertical posture. *Exp. Brain Res.* **1985**, *59*, 478–484. [CrossRef]
10. Slobounov, S.M.; Newell, K.M. Postural dynamics in upright and inverted stances. *J. Appl. Biomech.* **1996**, *12*, 185–196. [CrossRef]
11. Kochanowicz, A.; Niespodziński, B.; Marina, M.; Mieszkowski, J.; Biskup, L.; Kochanowicz, K. Relationship between postural control and muscle activity during a handstand in young and adult gymnasts. *Hum. Mov. Sci.* **2018**, *58*, 195–204. [CrossRef]
12. Kerwin, D.G.; Trewartha, G. Strategies for maintaining a handstand in the anterior-posterior direction. *Med. Sci. Sport. Exerc.* **2001**, *33*, 1182–1188. [CrossRef]
13. Brustio, P.; Rainoldi, A.; Petrigna, L.; Rabaglietti, E.; Pizzigalli, L. Postural stability during dual-and triple-task conditions: The effect of different levels of physical fitness in older adults. *Sci. Sport.* **2021**, *36*, 143–151. [CrossRef]
14. Croix, G.; Chollet, D.; Thouvarecq, R. Effect of expertise level on the perceptual characteristics of gymnasts. *J. Strength Cond. Res.* **2010**, *24*, 1458–1463. [CrossRef]
15. Trajano, G.S.; Taylor, J.L.; Orssatto, L.B.; McNulty, C.R.; Blazevich, A.J. Passive muscle stretching reduces estimates of persistent inward current strength in soleus motor units. *J. Exp. Biol.* **2020**, *223*, jeb229922. [CrossRef] [PubMed]
16. Thomas, E.; Martines, F.; Bianco, A.; Messina, G.; Giustino, V.; Zangla, D.; Iovane, A.; Palma, A. Decreased postural control in people with moderate hearing loss. *Medicine* **2018**, *97*, e0244. [CrossRef]
17. Barbero, M.; Merletti, R.; Rainoldi, A. *Atlas of Muscle Innervation Zones: Understanding Surface Electromyography and Its Applications*; Springer Science & Business Media: Berlin/Heidelberg, Germany, 2012.
18. Fritz, C.O.; Morris, P.E.; Richler, J.J. Effect size estimates: Current use, calculations, and interpretation. *J. Exp. Psychol. Gen.* **2012**, *141*, 2–18. [CrossRef]
19. Kerby, D.S. The simple difference formula: An approach to teaching nonparametric correlation. *Compr. Psychol.* **2014**, *3*, 11.IT.13.11. [CrossRef]
20. Omorczyk, J.; Bujas, P.; Puszczałowska-Lizis, E.; Biskup, L. Balance in handstand and postural stability in standing position in athletes practicing gymnastics. *Acta Bioeng. Biomech.* **2018**, *20*, 139–147.
21. Asseman, F.; Caron, O.; Crémieux, J. Is there a transfer of postural ability from specific to unspecific postures in elite gymnasts? *Neurosci. Lett.* **2004**, *358*, 83–86. [CrossRef]
22. Gautier, G.; Marin, L.; Leroy, D.; Thouvarecq, R. Dynamics of expertise level: Coordination in handstand. *Hum. Mov. Sci.* **2009**, *28*, 129–140. [CrossRef]
23. Brenière, Y.; Cuong Do, M.; Bouisset, S. Are dynamic phenomena prior to stepping essential to walking? *J. Mot. Behav.* **1987**, *19*, 62–76. [CrossRef] [PubMed]

24. Lepers, R.; Breniere, Y. The role of anticipatory postural adjustments and gravity in gait initiation. *Exp. Brain Res.* **1995**, *107*, 118–124. [CrossRef] [PubMed]
25. Yiou, E.; Caderby, T.; Delafontaine, A.; Fourcade, P.; Honeine, J.-L. Balance control during gait initiation: State-of-the-art and research perspectives. *World J. Orthop.* **2017**, *8*, 815. [CrossRef]
26. Halliday, S.E.; Winter, D.A.; Frank, J.S.; Patla, A.E.; Prince, F. The initiation of gait in young, elderly, and parkinson's disease subjects. *Gait Posture* **1998**, *8*, 8–14. [CrossRef]
27. Lu, C.; Huffmaster, S.L.A.; Harvey, J.C.; MacKinnon, C.D. Anticipatory postural adjustment patterns during gait initiation across the adult lifespan. *Gait Posture* **2017**, *57*, 182–187. [CrossRef] [PubMed]
28. Grabowiecki, M.; Rum, L.; Laudani, L.; Vannozzi, G. Biomechanical characteristics of handstand walking initiation. *Gait Posture* **2021**, *86*, 311–318. [CrossRef]
29. Rohleder, J.; Vogt, T. Teaching novices the handstand: A practical approach of different sport-specific feedback concepts on movement learning. *Sci. Gymnast. J.* **2018**, *10*, 29–42.
30. Hedbávný, P.; Sklenaříková, J.; Hupka, D.; Kalichová, M. Balancing in handstand on the floor. *Sci. Gymnast. J.* **2013**, *5*, 69–80.
31. Thomas, E.; Bianco, A.; Bellafiore, M.; Battaglia, G.; Paoli, A.; Palma, A. Determination of a strength index for upper body local endurance strength in sedentary individuals: A cross sectional analysis. *Springerplus* **2015**, *4*, 734. [CrossRef]
32. Mohammadi, M.; Yazici, A. A dynamic model for handstand in gymnastics. *IOSR-JSPE* **2016**, *3*, 8–11.
33. Runge, C.; Shupert, C.; Horak, F.; Zajac, F. Ankle and hip postural strategies defined by joint torques. *Gait Posture* **1999**, *10*, 161–170. [CrossRef] [PubMed]
34. Yamazaki, Y.; Suzuki, M.; Ohkuwa, T.; Itoh, H. Maintenance of upright standing posture during trunk rotation elicited by rapid and asymmetrical movements of the arms. *Brain Res. Bull.* **2005**, *67*, 30–39. [CrossRef] [PubMed]
35. Błaszczyszyn, M.; Szczęsna, A.; Piechota, K. Semg activation of the flexor muscles in the foot during balance tasks by young and older women: A pilot study. *Int. J. Environ. Res. Public Health* **2019**, *16*, 4307. [CrossRef]
36. Kochanowicz, A.; Niespodzinski, B.; Mieszkowski, J.; Marina, M.; Kochanowicz, K.; Zasada, M. Changes in the muscle activity of gymnasts during a handstand on various apparatus. *J. Strength Cond. Res.* **2019**, *33*, 1609–1618. [CrossRef]
37. Wyatt, H.E.; Vicinanza, D.; Newell, K.M.; Irwin, G.; Williams, G.K. Bidirectional causal control in the dynamics of handstand balance. *Sci. Rep.* **2021**, *11*, 1–9. [CrossRef] [PubMed]
38. Wyatt, H.; Vicinanza, D.; Newell, K.; Irwin, G.; Williams, G. Dynamics of Handstand Balance. In Proceedings of the 38th International Society of Biomechanics in Sport Conference, Liverpool, UK, 20–24 July 2020.
39. Puszczałowska-Lizis, E.; Omorczyk, J. The level of body balance in standing position and handstand in seniors athletes practicing artistic gymnastics. *Acta Bioeng. Biomech.* **2019**, *21*, 37–44. [PubMed]

Journal of
*Functional Morphology
and Kinesiology*

Review

Saddle Pressures Factors in Road and Off-Road Cyclists of Both Genders: A Narrative Review

Domenico Savio Salvatore Vicari [1], Antonino Patti [1], Valerio Giustino [1,*], Flavia Figlioli [1], Giuseppe Alamia [1], Antonio Palma [1,2] and Antonino Bianco [1]

[1] Sport and Exercise Sciences Research Unit, Department of Psychology, Educational Science and Human Movement, University of Palermo, 90141 Palermo, Italy

[2] Regional Sports School of CONI Sicilia, 90141 Palermo, Italy

* Correspondence: valerio.giustino@unipa.it

Abstract: The contact point of the pelvis with the saddle of the bicycle could generate abnormal pressure, which could lead to injuries to the perineum in cyclists. The aim of this review was to summarize in a narrative way the current literature on the saddle pressures and to present the factors that influence saddle pressures in order to prevent injury risk in road and off-road cyclists of both genders. We searched the PubMed database to identify English-language sources, using the following terms: "saddle pressures", "pressure mapping", "saddle design" AND "cycling". We also searched the bibliographies of the retrieved articles. Saddle pressures are influenced by factors such as sitting time on the bike, pedaling intensity, pedaling frequency, trunk and hand position, handlebars position, saddle design, saddle height, padded shorts, and gender. The jolts of the perineum on the saddle, especially on mountain bikes, generate intermittent pressures, which represent a risk factor for various pathologies of the urogenital system. This review highlights the importance of considering these factors that influence saddle pressures in order to prevent urogenital system injuries in cyclists.

Keywords: biomechanics; bicycle; cyclists; saddle pressure; perineal pressure; urogenital system; injury prevention

Citation: Vicari, D.S.S.; Patti, A.; Giustino, V.; Figlioli, F.; Alamia, G.; Palma, A.; Bianco, A. Saddle Pressures Factors in Road and Off-Road Cyclists of Both Genders: A Narrative Review. *J. Funct. Morphol. Kinesiol.* **2023**, *8*, 71. https://doi.org/10.3390/jfmk8020071

Academic Editor: Roland Van den Tillaar

Received: 30 March 2023
Revised: 19 May 2023
Accepted: 21 May 2023
Published: 25 May 2023

1. Introduction

Cycling is one of the most popular sports in the world and plays an important role in promoting public health, even for those who practice it as an amateur [1,2]. Indeed, cycling, in all its forms and in all its disciplines, is a sport that involves many participants, both at an amateur and professional level.

Cycling is practiced not only on the road, but also off-road and on the track. Four cycling disciplines are part of the Olympics: road cycling, mountain biking, track cycling, and BMX. Road cycling is probably the most widespread and followed discipline of cycling. The races can be one day or in stages. Cross country is the only Olympic discipline of mountain biking. The races take place on undulating mixed circuits with technical descents, forest roads, rock gardens, and obstacles (even artificial) from 4 to 6 km, and the route is repeated several times, depending on the travel time per lap, for a total time race of ≈120 min [3]. Track cycling features several races held within a velodrome, consisting of two straights and two steep curves. BMX is a cycling discipline that uses a small and resistant bike suitable for performing various stunts.

Very often, young people approach the bicycle with a mountain bike, attracted by the sense of freedom and escape from the daily chaos that derives from its use. In fact, the mountain bike, having wider and knobby tires and suspensions, offers the possibility of pedaling on uneven dirt paths and in close contact with nature.

It is important to note that a suitable position on the bike guarantees a good joint dynamic balance, correct breathing, excellent aerodynamics, and adequate weight distribution, favoring maximum muscle work without compromising the optimal balance of the physiology of the entire musculoskeletal system [4,5]. Bicycle fitting is the adjustment of the bicycle to the physical and performance requirements of the cyclist in order to meet the cyclist's goals and needs. An adequate bicycle fitting, performed by the bicycle fitter, can positively influence performance and the perception of comfort and reduce the risk of injury [4,6].

Indeed, regular cycling has an effect on the prevention of many chronic diseases (e.g., cardiovascular disease, diabetes, cancer, hypertension, obesity, depression, and osteoporosis) [2,7,8]. However, cycling can lead to injuries, most of which affect the urogenital system, causing genital numbness, erectile dysfunction (ED), priapism, infertility, and hematuria and affecting serum prostate-specific antigen (PSA) levels [9].

The man–bicycle combination is achieved through five points of contact, such as the two hands on the handlebars, the two feet on the pedals, and the pelvis on the saddle. Pressures are generated at these points of contact given by the distribution of the cyclist's weight. Saddle pressures are generated when the cyclist is seated, and, of course, no perineal pressure is generated when pedaling in standing position. In fact, the study by Sommer et al. (2001) [10] confirmed that pedaling in an upright position does not cause any alteration of the penile blood supply. As a matter of fact, the constant pressure exerted on the bicycle saddle could be the cause of non-traumatic injuries [11], ranging from saddle sores to more serious disorders related to the urogenital system [12].

Perineal discomfort or pain is very common on a bicycle and can be caused by various factors that can affect the normal distribution of pressure in the saddle. As a matter of fact, the saddle is recognized as the major extrinsic risk factor in the development of seat discomfort and perineal pathologies during cycling [13].

Analyzing the pressures in the saddle and controlling the factors that influence it can limit the risk of injury. In the last decade, the measurement of pressures in the saddle and the related compression on the perineum has been investigated [14]. A survey of 2,774 cyclists and 1,158 non-cyclists revealed that cycling is associated with a significantly higher risk of perineal numbness and urethral stricture development [12]. Up to 91% of cyclists experienced perineal numbness. The incidence of erectile dysfunction, especially in long-distance cyclists, was found to be 13%, which is significantly higher than in the general population [15]. Overall, 19% of cyclists who had a weekly distance of more than 400 km complained of erectile dysfunction. Moreover, perineal numbness was reported by 61% of cyclists.

In several sports, the analysis of the pressures in the different points of contact between the athlete and a surface is now widespread. For example, in running athletes, given the increasing attention to the relationship between shoes and running biomechanics [16,17], insole-based sensors that record pressures represent a methodology used to study foot–shoe interactions [18]. Similarly, other studies have investigated the pressure exerted on the saddle on horses and the correlation with back pain in riding horses [19]. Similarly, other studies investigated the differences in plantar pressure distribution between athletes of different sports [20] and between different technical gestures in athletes who practice the same sport [21], and the effects of training or technique [22].

Hence, the need to study saddle pressures in cyclists is of fundamental importance. This review aimed to summarize in a narrative way the current literature on the saddle pressures and the factors that influence saddle pressures in road and off-road cyclists of both genders.

2. Methods

For this review, we searched the PubMed database to identify English-language sources using the following terms: "saddle pressures", "pressure mapping", "saddle design" AND "cycling". We also searched the bibliographies of the retrieved articles. We have included only English-language articles, with no publication time limit.

3. Saddle Pressures Measurement

With technological progress, new instruments have been developed to optimize bicycle fitting. The measurement of pressures distribution in the saddle is essential because it analyzes the points of contact of the pelvis, where there are very delicate biological tissues, on the saddle. Saddle pressures are measured using an instrument applied directly to the saddle and equipped with sensors. The instrument allows the evaluation of the contact area, the distribution of the contact pressures, the distance between the ischial tuberosities, and the intermittence of the contact pressures. In detail, the contact area detects the surface of the saddle covered by the pelvis. The distribution of the contact pressures detects the region with a higher or lower load on the saddle. Among these parameters belong the vertical force (i.e., the integral of the pressure applied to the saddle area frame by frame and normalized to the cyclist's body weight), the mean pressure (i.e., the mean of the pressure values across all mask sensors), and the peak pressure (i.e., the highest value among the pressures recorded) [23]. Once the ischial tuberosities have been detected, the distance between the ischial tuberosities, i.e., how far apart they are from each other, is measured. The intermittence of the contact pressures represents the intermittent pressures exerted by the jerks of the pelvis on the saddle. In fact, the evaluation of the pressures in the saddle can take place in a static or dynamic mode. In static, saddle pressures are recorded with cyclists sitting on the saddle, but without pedaling [24]. In dynamic, the measurement of saddle pressures takes place while the cyclist pedals at a defined pedaling frequency and intensity.

4. High Saddle Pressures and Risk of Injury

In recent years, the scientific literature has focused on the analysis of the mechanical causes of overload injuries affecting the urogenital system in cyclists [23]. The incidence of bicycle-related urogenital symptoms varies [9]. The onset of some disorders is only occasional [25,26], while others, in particular those associated with perineal compression, are more frequent [9]. Sommer et al. (2001) reported an incidence of genital numbness of 61% and erectile dysfunction of 24% in male cyclists whose weekly training exceeded 400 km [10]. Few studies have analyzed the incidence of genital and pelvic floor symptoms in female cyclists [27]. Nonetheless, there is evidence that female cyclists suffer from similar problems as male cyclists, ranging from minor skin lesions to serious conditions, such as pain and neuropathies [27]. The cause of both genital numbness and erectile dysfunction appears to be compression of the pudendal nerve during pedaling [23]. Considering that excessive pressures leads to transient hypoxemia of the nerve [28], the duration of these compressions seems to be more relevant than the intensity of the pressure itself [9,29]. Several studies in the literature show that excessive pressures in the anterior region of the saddle are harmful to erectile tissues compared to pressures recorded in the posterior region of the saddle [30,31]. For this reason, reducing the compressive load on soft tissues is the main goal for the development of bicycle saddle geometries [32].

The factors that influence the distribution of pressures in the saddle are analyzed in the following paragraphs and are shown in Table 1.

Table 1. Factors influencing saddle pressure distributions.

First Author	Year	Measurements	Main Results
		Influence of pedaling on saddle pressures	
Carpes, F.P. [33]	2009	Influence of 2 different pedaling intensities (150 W and 300 W) on saddle pressures in male and female cyclists.	The mean pressure increased with increasing pedaling intensity in males. Furthermore, using the holed saddle, the mean pressure increased with increasing pedaling intensity, regardless of gender.

Table 1. *Cont.*

First Author	Year	Measurements	Main Results
Bressel, E. [34]	2005	Influence of different pedaling intensities (118 W and 300 ± 82.4 W) with different hand position (tops and drops) on saddle pressures in male and female cyclists.	The pressures decreased over most regions of the saddle at higher pedaling intensities and drop hand position.
Influence of saddle on saddle pressures			
Potter, J.J. [31]	2008	Influence of two saddle designs (gender-neutral and female-specific) with two different hand positions in female cyclists.	Using the female-specific saddle, the normalized maximum anterior pressure with drop hand position decreased compared to the gender-neutral saddle. No significant differences were found between saddles in the maximum pressure with tops hand position or in the posterior region with drop hand position.
Carpes, F.P. [33]	2009	Influence of 2 saddle designs (plain saddle and holed saddle) at 2 pedaling intensities (150 W and 300 W) in male and female cyclists.	The saddle design had no effect on pressures, with the only significant difference found when the peak pressure was compared between the holed saddle and the plain saddle during a pedaling intensity of 150 W in females.
Gámez, J. [35]	2008	Influence of three different saddle heights on comfort in cyclists (measuring saddle pressures).	The best level of comfort was the intermediate one; with this height, the peak pressure distribution and activation of lower limb muscles decreased.
Verma, R. [36]	2016	Influence of five different saddle positions (neutral, upward, downward, forward, and backward) on comfort in cyclists (measuring saddle pressures).	The discomfort increased with upward and backward positions compared to neutral. The minimum force became less negative with forward position compared to neutral position. The degree of variability of CoP increased in the upward and backward position.
Influence of handlebars on saddle pressures			
Partin, S.N. [37]	2012	Influence of three different handlebar-to-saddle heights (at saddle level, lower than saddle, or higher than saddle) on saddle pressures.	Low handlebar-to-saddle placement was associated with greater perineal saddle pressure.
Influence of trunk position and hand position on saddle pressures			
Carpes, F.P. [13]	2009	Effects of two trunk positions (upright and forward) on the pressures exerted in two saddles with different designs.	No significant differences in saddles pressures between the two trunk positions in females. The forward position of the trunk led to lower pressures using the saddle with holes in males.
Potter, J.J. [31]	2008	Influence of two hand positions (tops and drops) on saddle pressures.	From tops to drops hand position, the CoP moved forward, the normalized force and maximum pressure on the posterior region decreased. Only female cyclists showed an increased normalized force and maximum pressure in the anterior region.
Bressel, E. [38]	2003	Influence of hand position (tops and drops) on saddle pressures.	With the cyclist forward on the handlebars (drops position), there was an undesirable anterior pressure at the perineum.
Bressel, E. [34]	2005	Influence of hand position (tops and drops) on saddle pressures.	The drops hand position decreased saddle pressures in most regions.
Influence of padded shorts on saddle pressures			
Marcolin, G. [23]	2015	Influence of three padded shorts with different densities (basic, intermediate, resistance) on saddle pressures.	The vertical force and mean pressure on the saddle significantly decreased using the basic and endurance padded shorts. The peak pressure on the perineal area significantly increased only using the basic padded shorts.

Table 1. *Cont.*

First Author	Year	Measurements	Main Results
		Influence of sitting time on the bike on saddle pressures	
Bond, R.E. [39]	1975	Influence of sitting time or standing up on pedals on saddle pressures.	Often, pedaling standing up and taking breaks during pedaling are good habits to reduce saddle pressures and prevent the onset of urogenital pathologies.

Legend: W, Watt; CoP, Centre of Pressure.

5. Influence of Sitting Time on the Bike on Saddle Pressures

Cyclists, while pedaling, are not always sitting in the saddle. In fact, very often cyclists pedal standing up, and this factor is given by personal attitudes, the characteristics of the route, and the specialty practiced.

In fact, mountain bikers frequently find themselves in an upright position for very steep and bumpy climbs, where it is necessary to release greater power peaks to tackle them, and for very long descents, where the cyclist is not sitting in the saddle to overcome obstacles.

Conversely, road cyclists spend more time sitting in the saddle due to the regularity of their routes and the lack of obstacles to overcome. Professional road cyclists cover approximately 30,000 to 35,000 km per year between training and competition, and some races, such as the Tour de France, last 21 days (~100 h of racing), during which cyclists cover more than 3500 km [40].

Time spent sitting in the saddle could be a factor influencing saddle pressures and related urogenital pathologies. A few studies in the literature have analyzed saddle pressures for a long time in a sitting position. Often, pedaling standing up and taking breaks during the activity are good habits to reduce pressure on the saddle and prevent the onset of urogenital pathologies [39]. In this sense, off-road cyclists have an advantage over road cyclists because the discipline itself often invites them to pedal standing up.

It could be interesting to evaluate the pressure in the saddle over time to know the maximum time window in which it is possible to remain seated in the saddle while pedaling while maintaining comfort.

6. Influence of Pedaling Intensity on Saddle Pressures

Several studies measuring saddle pressures in a dynamic mode confirmed that pedaling intensity could be an influencing factor.

Carpes et al. (2009) [33] evaluated the effects of 2 different pedaling intensities (150 W and 300 W) and 2 saddle designs (plain saddle and holed saddle) on saddle pressure in male and female cyclists, showing that with the plain saddle, the mean pressure increased with increasing pedaling intensity in males. Furthermore, using the holed saddle, the mean pressure increased with increasing pedaling intensity, regardless of gender.

The study by Bressel et al. (2005) [34] investigated pressures at different pedaling intensities (118 W and 300 ± 82.4 W) and different hand positions (tops and drops) in male and female cyclists. The authors found decreased pressures over most regions of the saddle at higher pedaling intensities and the drop hand position. This may have occurred due to a partial weight shift from the saddle to the pedals as pedaling intensity is increased.

7. Influence of Pedaling Frequency on Saddle Pressures

Pedaling frequency could vary the pressure distribution in the saddle. A very high pedaling frequency could lead to higher pressure values in the saddle. In this case, it is advisable to change to a higher gear [39].

8. Influence of Trunk Position and Hand Position on Saddle Pressures

The position of the cyclist's trunk and hand could influence the pressure distribution in the saddle. Carpes et al. (2009) [13] analyzed the effects of trunk position on the pressures exerted in two saddles with different designs in recreational cyclists of both genders seated in a bicycle in static position. In the latter study, saddle pressures were measured on two saddle models (i.e., with and without holes) and in two trunk positions (i.e., upright and forward). The results of this study showed no significant differences in saddle pressures between the two trunk positions in females. Instead, a significant difference was found between trunk positions using the saddle with holes in males. In detail, the forward position of the trunk led to lower pressure using the saddle with holes.

In contrast, the study by Potter et al. (2008) [31] that investigated the influence of gender and hand position on saddle pressures showed that when cyclists pedal and move hand position from the tops to drops, the centers of pressure in all regions move forward, the normalized force and maximum pressure on the posterior region decrease, and female cyclists show an increase in normalized force and maximum pressure in the anterior region.

Bressel et al. (2003) [38] showed an undesirable pressure to the anterior perineum when the cyclist leans forward on the handlebar of the bicycle.

In another study by Bressel et al. (2005) [34], the authors found that the drop hand position decreased saddle pressures in most regions. This could have occurred due to a partial shift of the cyclists' weight from the saddle to the handlebars as a result of moving from the top to drop the position of the hands on the handlebars.

However, it has been suggested that body weight bearing on the saddle at the level of the ischial tuberosities is an effective strategy to minimize pressure on the perineum [32].

9. Influence of Handlebars Position on Saddle Pressures

In the study by Partin et al. (2012) [37], it was demonstrated that the position of the handlebars can influence the pressures in the saddle of female cyclists. Three different handlebar-to-saddle heights were evaluated (at saddle level, lower than saddle, or higher than saddle), and low handlebar-to-saddle placement was associated with greater perineal saddle pressure.

10. Influence of Saddle Design on Saddle Pressures

Several studies have investigated the effects of saddle type on saddle pressures. Indeed, the saddle appears to be the main risk factor for urogenital problems. This aspect has contributed to the development of saddles with new designs.

Potter et al. (2008) [31] evaluated the influence of two saddle designs (saddle A, i.e., gender-neutral; saddle B, i.e., female-specific) by adopting two different hand positions in female cyclists. Using saddle B showed a reduction in the normalized maximum anterior pressure with drop hand position compared to saddle A. No significant differences were found between saddles in the maximum pressure with tops hand position or in the posterior region with drop hand position.

The study by Carpes et al. (2009) [33] investigated the influence on pressure in 2 saddles with different designs (plain saddle and holed saddle) at 2 pedaling intensities (150 W and 300 W) in male and female cyclists. The results of this study showed that the saddle design had no effect on pressures, with the only significant difference found when the peak pressure was compared between the holed saddle and the plain saddle during a pedaling intensity of 150 W in females. Higher pressures can be caused by contact with small surfaces on the edge of the saddle hole. On the plain saddle, the pressures were distributed over the entire surface, while on the holed saddle, there were points of higher pressure, which were given by the ischial tuberosities. This last aspect indicates that the contact of the body weight was mainly on the ischial tuberosities, reducing the pressures on the perineal area.

11. Influence of Saddle Height on Saddle Pressures

The optimal saddle height is an important aspect for a functional bike fitting [4]. The effects of changes in saddle height on the influence of perineal pressures and discomfort and pain have not been extensively studied. The study by De Looze et al. (2003) [41]

demonstrated that pressure distribution and muscle activation are parameters related to discomfort. Setting an appropriate saddle height is very important to increase comfort, prevent injuries, make pedaling more efficient, and improve performance [36,42]. The scientific literature reports different methods to adjust the height of the saddle [42]. Some of these take into account the anthropometric measurements of the cyclist, and others use 2D or 3D kinematic analysis evaluating the range of the joint angle [43]. Christiaans et al. (1998) [5] suggested that the best saddle height for comfort was 106% of the inseam height in men and 107% of the inseam height in women. Another study by Scoz et al. (2022) [43] evaluated subjects with a standardized bicycle fitting procedure based on 3D kinematic data. The results of this study demonstrated that fitting across 15 angular ranges (minimum ankle, maximal ankle, ankle range, ankle angle at bottom, maximum knee flexion, maximum knee extension, knee angle range, knee forward of foot, knee forward of spindle, knee travel tilt, knee lateral tilt, knee lateral travel, hip angle closed, hip angle open, hip angle range, hip lateral travel, back angle, shoulder angle to wrist, shoulder angle to elbow) was sufficient to produce large, long-term effects on pain, fatigue, and comfort. The study by Gamez et al. (2008) investigated the influence of 3 different saddle heights (i.e., 684.1 mm, 709.1 mm, 734.1 mm) on comfort in cyclists. The authors found that the 709.1 mm high saddle was best because the peak pressure distribution and lower-limb muscle activation decreased [35]. The study by Verma et al. (2016) [36] investigated the influence of different saddle positions (neutral, upward, downward, forward, backward) on saddle pressures distribution. The authors found that the discomfort increased with the upward and backward positions compared to neutral. The minimum force became less negative with the forward position compared to the neutral position. The degree of variability of CoP increased in the upward and backward positions.

A recent work by Wang et al. (2020) showed that low saddle height is related to increased knee adduction moments with longer duration in recreational cyclists [44].

Considering that a higher-than-normal saddle height could generate higher pressures, it is very important to adjust the saddle height and the backwardness of the saddle to prevent urogenital pathologies and ensure greater comfort for cyclists.

12. Influence of Padded Shorts on Saddle Pressures

A recent study evaluated the influence of various cycling padded shorts on saddle pressures in order to preserve the perineal region and increase the level of comfort [23].

A total of 3 padded shorts with different densities were evaluated: basic (60 kg/m^3), intermediate (80 kg/m^3), and resistance (80 and 120 kg/m^3). Saddle pressures were recorded during three trials of twenty minutes each, in which cyclists wore one of the three padded shorts for each trial. The authors found that the vertical force and the mean pressure on the saddle significantly decreased using the basic and endurance padded shorts. Moreover, the peak pressure on the corresponding perineal area of cyclists was significantly increased only with the use of the basic padded shorts. Further, the increase in the length of the center of pressure of the 3 padded shorts at the end of the test suggested that the basic material was starting to lose its elastic properties due to its low foam density (60 kg/m^3). This study, in line with others in the literature, suggests that it is important to consider cyclists' padded shorts as a factor influencing saddle pressures [23,45].

13. Muscle Activity and Saddle Pressures

Several studies in the literature have evaluated muscle activity to improve pedaling economy and efficiency [46–50]. The major muscles typically studied were the gluteus maximus (GM), rectus femoris (RF), vastus lateralis (VL), vastus medialis (VM), biceps femoris (BF, long head), medial gastrocnemius (MG), and tibialis anterior (TA) [48]. The study by McDonald et al. (2021) [51] evaluated the muscle activity of these muscles and the changes in the saddle pressure mapping indexes with alteration of the effective seat tube angle (ESTA). To vary the latter, the handlebars and the saddle were moved forwards or backwards simultaneously by 3 cm. The main finding concerns the muscle activity of the BF, GM, and MG, which increased progressively from the forwards to the backwards position. In contrast, the muscle activity of TA decreased progressively from the forwards to the backwards position. No changes were found in the muscle activity of VM, VL, and RF across the different positions. Regarding saddle pressures, a decrease in the percentage of frontal versus rear pressure was found when participants moved from the backwards to the forwards saddle position. A decrease in the mean pubic pressure was also found from the backwards to the forwards position. The results of this study suggest that cyclists in the back position had greater pelvic rotation, therefore increasing perineal pressure and anterior/posterior pressure distribution to compensate. The results were opposite when the saddle was moved forward.

14. Gender Differences on Saddle Pressures

The scientific literature shows several studies that have investigated the difference between male and female cyclists in different aspects. The difference in pelvic geometry, power output, and bike fit can influence saddle pressures. For example, the greater width between the ischial tuberosities in female cyclists could reduce the load on the posterior bone structures and increase the load on the perineal region [31].

The study by Potter et al. (2008) [31] analyzed gender differences in the distribution of pressures in the saddle. The main aim was to investigate the influence of gender, power, and hand position on saddle pressure distribution. The results confirmed a gender difference in saddle pressures. In detail, with the hand in drop position, only female cyclists showed an increase in the normalized force and maximum pressure in the anterior region.

Similarly, Carpes et al. (2009) also found significant differences between genders with the same workload.

Bressel et al. (2005) [34] showed that females, compared with males, did not show lower peak total, posterior, left, and right pressures in the drop position.

An interesting paper by Hermans et al. (2016) [52] explored the prevalence and duration of urogenital overuse injuries and sexual dysfunctions in female cyclists, showing that after at least 2 h of cycling, dysuria (8.8%), stranguria (22.2%), genital numbness (34.9%), and vulvar discomfort (40.0%) were found.

Coutant-Foulc et al. (2014) [53] reported that female cyclists cycling long distances each week resulted in unilateral swelling of the labium majus.

These findings suggest that it is essential to take into account the gender of cyclists in changes in saddle pressures to prevent perineal injuries.

15. Road vs. Off-Road Cyclists: Difference in Saddle Pressures

Between the road bike and the mountain bike, there are substantial differences, both structural and related to the sport, that involve a difference in the variation of saddle pressures. The mountain bike does not allow different grips, such as with the road bike. For this reason, saddle pressures in mountain bike cyclists cannot be influenced by different grips. However, the off-road rough trails cause percussions in the cyclist's perineum. Genital numbness and erectile dysfunction can result from repeated perineal impacts on the bicycle saddle (micro-trauma).

Sanford et al. (2018) [24] evaluated the relationship between oscillation forces and perineal pressures among cyclists in a simulated laboratory setting by demonstrating a strong linear relationship between oscillation magnitude and perineal pressure during pedaling, mitigated by a saddle post shock absorber.

Studies in the scientific literature report that mountain bikers have a high frequency of extra-testicular and testicular disorders, associated with clinical symptoms in half of the bikers. These may be associated with a high rate of repeated micro-traumas of the scrotal contents [54]. Indeed, mountain bikers are at a higher risk of scrotal disorders than road cyclists [55]. Indeed, a prospective cohort study by Dettori et al. (2004) [56], aimed at exploring the occurrence of erectile dysfunction when using a mountain bicycle/road bicycle after a long-distance cycling event, showed that riding a mountain bicycle increases the risk of erectile dysfunction compared to a road bicycle on a long-distance ride.

Mitterberger et al. (2008) [55] investigated the risk of scrotal disorders in mountain bikers and on-road cyclists. The results showed that 94% of mountain bikers and 48% of on-road cyclists presented with abnormal findings on a scrotal ultrasound.

To the best of our knowledge, there are very few studies that have analyzed saddle pressures during outdoor mountain biking. In mountain biking, there are many factors that can influence the pressure in the saddle. Factors such as tire pressure, shock pressure, shock type, number of shock absorbers, terrain, climbing, and saddle contact time must be considered. Some of these factors are difficult to evaluate in the laboratory on cycle simulators, which is why it is interesting to evaluate the evolution of pressure in the saddle during outdoor pedaling on a mountain bike.

Few studies have so far analyzed the correlation between pressures in the different contact points, and knowing the pressures on the handlebars, in the saddle, and in the soles of the shoes could give us the possibility to identify postural anomalies of the cyclist and inadequate or inefficient pedaling techniques and to develop protocols to make man–bicycle interactions efficient and comfortable.

16. Conclusions

The analysis of saddle pressures is of fundamental importance to prevent the onset of urogenital pathologies in cyclists and to guarantee them a greater perception of comfort. The pressures under the perineal area of cyclists are harmful, and the main objective of bike fitting is to limit them as much as possible. This review suggests the importance of knowing the different factors that can influence the distribution of pressures in the saddle. Among these, technicians should be taking into account the influence of (a) pedaling intensity and frequency, (b) trunk position and hand position, (c) handlebars position, (d) saddle design and saddle height during a bike fit for each cyclist. These factors are individual and depend, among others, on gender, anthropometric measurements, and the type of bike. Moreover, they should recommend the most suitable type of padded shorts to use and give advice, such as getting up from the saddle while pedaling in the case of long periods of sitting time.

Considering which factors influence pressure in the saddle is useful for cycling practitioners. In particular, the use of devices capable of recording pressure in the saddle by technicians could make pedaling more efficient and comfortable. In fact, limiting pressure in the pubic region is the main objective to ensure adequate comfort for cyclists.

All the variables that can influence pressure in the saddle analyzed in this review represent risk factors for urogenital pathologies in cyclists. When fitting a cyclist's bike, these factors should be analyzed, taking into consideration other parameters, such as the weight, age, and level of the cyclist.

Professional bike fitters, cyclists, and trainers can use these findings to improve pedaling [43].

Author Contributions: Conceptualization, V.G. and A.B.; methodology, D.S.S.V., A.P. (Antonino Patti), and V.G.; writing—original draft preparation, D.S.S.V.; writing—review and editing, V.G.; visualization, G.A. and F.F.; supervision, A.P. (Antonio Palma) and A.B. All authors have read and agreed to the published version of the manuscript.

Funding: This research received no external funding.

Institutional Review Board Statement: Not applicable.

Informed Consent Statement: Not applicable.

Data Availability Statement: Not applicable.

Conflicts of Interest: The authors declare no conflict of interest.

References

1. Bassett, D.R., Jr.; Pucher, J.; Buehler, R.; Thompson, D.L.; Crouter, S.E. Walking, cycling, and obesity rates in Europe, North America, and Australia. *J. Phys. Act. Health* **2008**, *5*, 795–814. [CrossRef] [PubMed]
2. Oja, P.; Titze, S.; Bauman, A.; de Geus, B.; Krenn, P.; Reger-Nash, B.; Kohlberger, T. Health benefits of cycling: A systematic review. *Scand. J. Med. Sci. Sports* **2011**, *21*, 496–509. [CrossRef] [PubMed]
3. Impellizzeri, F.M.; Marcora, S.M. The physiology of mountain biking. *Sports Med.* **2007**, *37*, 59–71. [CrossRef] [PubMed]
4. Swart, J.; Holliday, W. Cycling Biomechanics Optimization—The (R) Evolution of Bicycle Fitting. *Curr. Sports Med. Rep.* **2019**, *18*, 490–496. [CrossRef] [PubMed]
5. Christiaans, H.H.; Bremner, A. Comfort on bicycles and the validity of a commercial bicycle fitting system. *Appl. Ergon.* **1998**, *29*, 201–211. [CrossRef]
6. Quesada, J.I.P.; Kerr, Z.Y.; Bertucci, W.M.; Carpes, F.P. The association of bike fitting with injury, comfort, and pain during cycling: An international retrospective survey. *Eur. J. Sport. Sci.* **2019**, *19*, 842–849. [CrossRef]
7. Warburton, D.E.; Nicol, C.W.; Bredin, S.S. Health benefits of physical activity: The evidence. *CMAJ* **2006**, *174*, 801–809. [CrossRef]
8. Hillman, M. Cycling offers important health benefits and should be encouraged. *BMJ* **1997**, *315*, 490. [CrossRef]
9. Leibovitch, I.; Mor, Y. The vicious cycling: Bicycling related urogenital disorders. *Eur. Urol.* **2005**, *47*, 277–286. [CrossRef]
10. Sommer, F.; Konig, D.; Graf, C.; Schwarzer, U.; Bertram, C.; Klotz, T.; Engelmann, U. Impotence and genital numbness in cyclists. *Int. J. Sports Med.* **2001**, *22*, 410–413. [CrossRef]
11. Dettori, N.J.; Norvell, D.C. Non-traumatic bicycle injuries: A review of the literature. *Sports Med.* **2006**, *36*, 7–18. [CrossRef] [PubMed]
12. Litwinowicz, K.; Choroszy, M.; Wrobel, A. Strategies for Reducing the Impact of Cycling on the Perineum in Healthy Males: Systematic Review and Meta-analysis. *Sports Med.* **2021**, *51*, 275–287. [CrossRef] [PubMed]
13. Carpes, F.P.; Dagnese, F.; Kleinpaul, J.F.; Martins, E.D.A.; Mota, C.B. Bicycle saddle pressure: Effects of trunk position and saddle design on healthy subjects. *Urol. Int.* **2009**, *82*, 8–11. [CrossRef]
14. Spears, I.R.; Cummins, N.K.; Brenchley, Z.; Donohue, C.; Turnbull, C.; Burton, S.; Macho, G.A. The effect of saddle design on stresses in the perineum during cycling. *Med. Sci. Sports Exerc.* **2003**, *35*, 1620–1625. [CrossRef] [PubMed]
15. Breda, G.; Piazza, N.; Bernardi, V.; Lunardon, E.; Caruso, A. Development of a new geometric bicycle saddle for the maintenance of genital-perineal vascular perfusion. *J. Sex. Med.* **2005**, *2*, 605–611. [CrossRef]
16. Wiegerinck, J.I.; Boyd, J.; Yoder, J.C.; Abbey, A.N.; Nunley, J.A.; Queen, R.M. Differences in plantar loading between training shoes and racing flats at a self-selected running speed. *Gait Posture* **2009**, *29*, 514–519. [CrossRef] [PubMed]
17. Willy, R.W.; Davis, I.S. Kinematic and kinetic comparison of running in standard and minimalist shoes. *Med. Sci. Sports Exerc.* **2014**, *46*, 318–323. [CrossRef] [PubMed]
18. Dixon, S.J. Use of pressure insoles to compare in-shoe loading for modern running shoes. *Ergonomics* **2008**, *51*, 1503–1514. [CrossRef]
19. Dittmann, M.T.; Arpagaus, S.; Hungerbuhler, V.; Weishaupt, M.A.; Latif, S.N. "Feel the Force"—Prevalence of Subjectively Assessed Saddle Fit Problems in Swiss Riding Horses and Their Association With Saddle Pressure Measurements and Back Pain. *J. Equine Vet. Sci.* **2021**, *99*, 103388. [CrossRef]
20. Feka, K.; Pomara, F.; Russo, G.; Piccione, M.C.; Petrucci, M.; Giustino, V.; Messina, G.; Iovane, A.; Palma, A.; Bianco, A. How do sports affect static baropodometry? An observational study among women living in southern Italy. *Hum. Mov.* **2019**, *20*, 9–16. [CrossRef]
21. Lambrich, J.; Muehlbauer, T. Effect of stroke direction on plantar pressure in each foot during the forehand and backhand stroke among healthy adult tennis players of different performance levels. *BMC Sports Sci. Med. Rehabil.* **2023**, *15*, 23. [CrossRef] [PubMed]
22. Bonaventura, R.E.; Giustino, V.; Chiaramonte, G.; Giustiniani, A.; Smirni, D.; Battaglia, G.; Messina, G.; Oliveri, M. Investigating prismatic adaptation effects in handgrip strength and in plantar pressure in healthy subjects. *Gait Posture* **2020**, *76*, 264–269. [CrossRef] [PubMed]
23. Marcolin, G.; Petrone, N.; Reggiani, C.; Panizzolo, F.A.; Paoli, A. Biomechanical Comparison of Shorts With Different Pads: An Insight into the Perineum Protection Issue. *Medicine* **2015**, *94*, e1186. [CrossRef] [PubMed]
24. Sanford, T.; Gadzinski, A.J.; Gaither, T.; Osterberg, E.C.; Murphy, G.P.; Carroll, P.R.; Breyer, B.N. Effect of Oscillation on Perineal Pressure in Cyclists: Implications for Micro-Trauma. *Sex. Med.* **2018**, *6*, 239–247. [CrossRef] [PubMed]
25. Somerhausen, N.D.S.A.; Geurde, B.; Couvreur, Y. Perineal nodular induration: The 'third testicle of the cyclist', an under-recognized pseudotumour. *Histopathology* **2003**, *42*, 615–616. [CrossRef]
26. Solomon, S.; Cappa, K.G. Impotence and bicycling. A seldom-reported connection. *Postgrad. Med.* **1987**, *81*, 99–102. [CrossRef]
27. Trofaier, M.L.; Schneidinger, C.; Marschalek, J.; Hanzal, E.; Umek, W. Pelvic floor symptoms in female cyclists and possible remedies: A narrative review. *Int. Urogynecol. J.* **2016**, *27*, 513–519. [CrossRef]
28. Guiotto, A.; Spolaor, F.; Albani, G.; Sawacha, Z. Could Proprioceptive Stimuli Change Saddle Pressure on Male Cyclists during Different Hand Positions? An Exploratory Study of the Effect of the Equistasi® Device. *Sports* **2022**, *10*, 88. [CrossRef]
29. Sommer, F.; Goldstein, I.; Korda, J.B. Bicycle riding and erectile dysfunction: A review. *J. Sex. Med.* **2010**, *7*, 2346–2358. [CrossRef]
30. Schrader, S.M.; Breitenstein, M.J.; Clark, J.C.; Lowe, B.D.; Turner, T.W. Nocturnal penile tumescence and rigidity testing in bicycling patrol officers. *J. Androl.* **2002**, *23*, 927–934.

31. Potter, J.J.; Sauer, J.L.; Weisshaar, C.L.; Thelen, D.G.; Ploeg, H.L. Gender differences in bicycle saddle pressure distribution during seated cycling. *Med. Sci. Sports Exerc.* **2008**, *40*, 1126–1134. [CrossRef] [PubMed]

32. Lowe, B.D.; Schrader, S.M.; Breitenstein, M.J. Effect of bicycle saddle designs on the pressure to the perineum of the bicyclist. *Med. Sci. Sports Exerc.* **2004**, *36*, 1055–1062. [CrossRef] [PubMed]

33. Carpes, F.P.; Dagnese, F.; Kleinpaul, J.F.; Martins, E.A.; Mota, C.B. Effects of workload on seat pressure while cycling with two different saddles. *J. Sex. Med.* **2009**, *6*, 2728–2735. [CrossRef] [PubMed]

34. Bressel, E.; Cronin, J. Bicycle seat interface pressure: Reliability, validity, and influence of hand position and workload. *J. Biomech.* **2005**, *38*, 1325–1331. [CrossRef] [PubMed]

35. Gámez, J.; Zarzoso, M.; Raventós, A.; Valero, M.; Alcántara, E.; López, A.; Prat, J.; Vera, P. Determination of the optimal saddle height for leisure cycling (P188). *Eng. Sport 7* **2008**, 255–260.

36. Verma, R.; Hansen, E.A.; de Zee, M.; Madeleine, P. Effect of seat positions on discomfort, muscle activation, pressure distribution and pedal force during cycling. *J. Electromyogr. Kinesiol.* **2016**, *27*, 78–86. [CrossRef]

37. Partin, S.N.; Connell, K.A.; Schrader, S.; LaCombe, J.; Lowe, B.; Sweeney, A.; Reutman, S.; Wang, A.; Toennis, C.; Melman, A.; et al. The bar sinister: Does handlebar level damage the pelvic floor in female cyclists? *J. Sex. Med.* **2012**, *9*, 1367–1373. [CrossRef]

38. Bressel, E.; Larson, B.J. Bicycle seat designs and their effect on pelvic angle, trunk angle, and comfort. *Med. Sci. Sports Exerc.* **2003**, *35*, 327–332. [CrossRef]

39. Bond, R.E. Distance Bicycling May Cause Ischemic Neuropathy of Penis. *Phys. Sportsmed.* **1975**, *3*, 54–56. [CrossRef]

40. Lucia, A.; Hoyos, J.; Chicharro, J.L. Physiology of professional road cycling. *Sports Med.* **2001**, *31*, 325–337. [CrossRef]

41. de Looze, M.P.; Kuijt-Evers, L.F.; van Dieen, J. Sitting comfort and discomfort and the relationships with objective measures. *Ergonomics* **2003**, *46*, 985–997. [CrossRef] [PubMed]

42. Bini, R.; Hume, P.A.; Croft, J.L. Effects of bicycle saddle height on knee injury risk and cycling performance. *Sports Med.* **2011**, *41*, 463–476. [CrossRef] [PubMed]

43. Scoz, R.D.; de Oliveira, P.R.; Santos, C.S.; Pinto, J.R.; Melo-Silva, C.A.; de Judice, A.F.T.; Mendes, J.J.B.; Ferreira, L.M.A.; Amorim, C.F. Long-Term Effects of a Kinematic Bikefitting Method on Pain, Comfort, and Fatigue: A Prospective Cohort Study. *Int. J. Environ. Res. Public Health* **2022**, *19*, 12949. [CrossRef] [PubMed]

44. Wang, Y.; Liang, L.; Wang, D.; Tang, Y.; Wu, X.; Li, L.; Liu, Y. Cycling with Low Saddle Height is Related to Increased Knee Adduction Moments in Healthy Recreational Cyclists. *Eur. J. Sport Sci.* **2020**, *20*, 461–467. [CrossRef] [PubMed]

45. Mellion, M.B. Common cycling injuries. Management and prevention. *Sports Med.* **1991**, *11*, 52–70. [CrossRef]

46. Baum, B.S.; Li, L. Lower extremity muscle activities during cycling are influenced by load and frequency. *J. Electromyogr. Kinesiol.* **2003**, *13*, 181–190. [CrossRef]

47. Dorel, S.; Drouet, J.M.; Couturier, A.; Champoux, Y.; Hug, F. Changes of pedaling technique and muscle coordination during an exhaustive exercise. *Med. Sci. Sports Exerc.* **2009**, *41*, 1277–1286. [CrossRef]

48. Hug, F.; Dorel, S. Electromyographic analysis of pedaling: A review. *J. Electromyogr. Kinesiol.* **2009**, *19*, 182–198. [CrossRef]

49. O'Bryan, S.J.; Brown, N.A.; Billaut, F.; Rouffet, D.M. Changes in muscle coordination and power output during sprint cycling. *Neurosci. Lett.* **2014**, *576*, 11–16. [CrossRef]

50. Saunders, M.J.; Evans, E.M.; Arngrimsson, S.A.; Allison, J.D.; Warren, G.L.; Cureton, K.J. Muscle activation and the slow component rise in oxygen uptake during cycling. *Med. Sci. Sports Exerc.* **2000**, *32*, 2040–2045. [CrossRef]

51. McDonald, R.; Holliday, W.; Swart, J. Muscle recruitment patterns and saddle pressure indexes with alterations in effective seat tube angle. *Sports Med. Health Sci.* **2022**, *4*, 29–37. [CrossRef] [PubMed]

52. Hermans, T.J.; Wijn, R.P.; Winkens, B.; Van Kerrebroeck, P.E. Urogenital and Sexual Complaints in Female Club Cyclists—A Cross-Sectional Study. *J. Sex. Med.* **2016**, *13*, 40–45. [CrossRef] [PubMed]

53. Coutant-Foulc, P.; Lewis, F.M.; Berville, S.; Janssen, B.; Guihard, P.; Renaut, J.J.; Plantier, F.; Calonje, E.; Moyal-Barracco, M. Unilateral vulval swelling in cyclists: A report of 8 cases. *J. Low. Genit. Tract Dis.* **2014**, *18*, e84–e89. [CrossRef] [PubMed]

54. Frauscher, F.; Klauser, A.; Hobisch, A.; Pallwein, L.; Stenzl, A. Subclinical microtraumatisation of the scrotal contents in extreme mountain biking. *Lancet* **2000**, *356*, 1414. [CrossRef] [PubMed]

55. Mitterberger, M.; Pinggera, G.M.; Neuwirt, H.; Colleselli, D.; Pelzer, A.; Bartsch, G.; Strasser, H.; Gradl, J.; Pallwein, L.; Frauscher, F. Do mountain bikers have a higher risk of scrotal disorders than on-road cyclists? *Clin. J. Sport Med.* **2008**, *18*, 49–54. [CrossRef] [PubMed]

56. Dettori, J.R.; Koepsell, T.D.; Cummings, P.; Corman, J.M. Erectile dysfunction after a long-distance cycling event: Associations with bicycle characteristics. *J. Urol.* **2004**, *172*, 637–641. [CrossRef]

Journal of
*Functional Morphology
and Kinesiology*

Article

Elastic Taping Application on the Neck: Immediate and Short-Term Impacts on Pain and Mobility of Cervical Spine

Luca Russo [1], Tommaso Panessa [2], Paolo Bartolucci [1], Andrea Raggi [3], Gian Mario Migliaccio [4,*], Alin Larion [5,†] and Johnny Padulo [6,†]

[1] Department of Human Sciences, Università Telematica Degli Studi IUL, 50122 Florence, Italy; l.russo@iuline.it (L.R.); p.bartolucci@iuline.it (P.B.)
[2] Department of Biotechnological and Applied Clinical Sciences, University Degli Studi dell'Aquila, 67100 L'Aquila, Italy; tommaso_panessa@libero.it
[3] Laboratory of Biomechanics, FGP srl, 37062 Verona, Italy; andrea@fgpsrl.it
[4] Department of Human Science and Promotion of Quality of Life, San Raffaele University, 00166 Rome, Italy
[5] Faculty of Physical Education and Sport, Ovidius University of Constanta, 900029 Constanta, Romania; alinlarion@yahoo.com
[6] Department of Biomedical Sciences for Health, Università degli Studi di Milano, 20133 Milan, Italy; johnny.padulo@unimi.it
*   Correspondence: gianmario.migliaccio@uniroma5.it
†   These authors contributed equally to this work.

Abstract: The aim of this study was to measure the effects on three-planar active cervical range of motion (ACROM) and self-perceived pain of elastic taping (ET) application in the cervical area. Thirty participants (*n:* 22-M and 8-F, age 35.4 ± 4.4 years; body height 173.1 ± 8.4 cm; body mass 73.5 ± 12.8 kg) in the study group (SG) and twenty participants (*n:* 11-M and 9-F, age 32.6 ± 3.9 years; body height 174.9 ± 10.9 cm; body mass 71.2 ± 12.9 kg) in the control group (CG) were recruited. All subjects had neck and cervical pain in baseline condition. Each group performed an ACROM test and measured the perceived pain in the neck based on the Numerical Rating Scale (NRS 0–10, a.u.) at the baseline (T0), after 20′ from the ET application (T1), and after three days of wearing the ET application (T2). Between T0 and T1, an ET was applied to the cervical area of the SG participants. Statistical analysis did not show any significant change in CG in any measurement session for ACROM and neck pain parameters. Conversely, the SG showed significant improvements for ACROM rotation to the left (T0 64.8 ± 7.7°–T2 76.0 ± 11.1° $p < 0.000$) and right (T0 66.0 ± 11.9°–T2 74.2 ± 9.6° $p < 0.000$), lateral inclination to the left (T0 37.5 ± 6.9°–T2 40.6 ± 10.8° $p < 0.000$) and right (T0 36.5 ± 7.9°–T2 40.9 ± 5.2° $p < 0.000$), extension (T0 47.0 ± 12.9°–T2 55.1 ± 12.3° $p < 0.001$), and flexion (T0 55.0 ± 3.6°–T2 62.9 ± 12.0° $p < 0.006$). A significant decrease was also measured in SG for pain NRS between T0 and T2 (T0 7.5 ± 1.0°–T1 5.5 ± 1.4–T2 1.4 ± 1.5° $p < 0.000$). In conclusion, a bilateral and symmetrical ET cervical application is useful to enhance multiplanar ACROM and reduce subjective self-perceived cervical pain when it is needed. Based on the evidence, the use of ET on the neck is recommended for managing neck motion restrictions and pain in adult individuals.

Keywords: cervical ROM; elastic taping; neck pain; kinesiology; musculoskeletal health

Citation: Russo, L.; Panessa, T.; Bartolucci, P.; Raggi, A.; Migliaccio, G.M.; Larion, A.; Padulo, J. Elastic Taping Application on the Neck: Immediate and Short-Term Impacts on Pain and Mobility of Cervical Spine. *J. Funct. Morphol. Kinesiol.* **2023**, *8*, 156. https://doi.org/10.3390/jfmk8040156

Academic Editor: William R. Taylor

Received: 12 September 2023
Revised: 24 October 2023
Accepted: 1 November 2023
Published: 7 November 2023

1. Introduction

The cervical spine, consisting of seven vertebrae, is known for its exceptional mobility within the spinal column. It plays a crucial role in facilitating multiplanar movements of the head in space. Specifically, the motion of the cervical spine and head complex can be categorized into right–left rotation, right–left lateral inclination, flexion, and extension, meaning forward and backward head motion, respectively [1]. These movements can be performed individually on a single anatomical plane or via combined multiplanar actions, depending on the functional demands of the environment.

Given these unique characteristics, preserving or restoring the physiological active range of motion (ACROM) of the cervical spine becomes a significant challenge in daily life, particularly in sports and various activities, as it contributes to maintaining optimal health [2,3]. Numerous factors can impede and restrict the ACROM, including traumatic injuries, repetitive movements, occupational practices, prolonged postures, head positioning, and other factors [4–8]. Diminished ACROM can have various detrimental effects on the musculoskeletal system, affecting the use of the eyes, other spinal segments, shoulder mobility, etc. Additionally, a reduced ACROM is a common observation among individuals experiencing neck pain [9–13]. In light of this context, two primary considerations arise: (1) the need for a comprehensive evaluation of the ACROM and (2) the implementation of effective strategies to enhance it.

Regarding the ACROM assessment, professionals have the option to conduct tests with or without devices [14]. Qualitative analyses without devices are cost effective and straightforward, but there is no change to compare data over time [15]. Within the realm of instrument-based evaluation, various solutions exist, including video analysis and 3D motion capture systems. However, these methods are more suited to analyzing posture and complex movements in specific contexts [16–18]. As a result, the most utilized and practical approach involves the use of inertial sensors [3,19–22].

These tools enable professionals to accurately measure the ACROM in a convenient and efficient manner, without causing discomfort to the subjects. Typically, an inertial sensor is positioned on the subject's forehead and moves in conjunction with the head, capturing angles and the range of motion during neck active movements [19–22], avoiding any compensatory movement of the shoulder girdle.

Regarding strategies for preserving and improving ACROM, multiple approaches can be considered, such as reducing screen time, minimizing sedentary behavior, receiving massages, engaging in targeted exercises, utilizing cupping techniques, and employing elastic taping [23–27]. Among these solutions, elastic taping (ET) possesses a distinctive characteristic: it can be worn directly on the skin for an extended period, continuously working 24 h a day and adapting to the subject's movements until its removal. This unique characteristic makes ET a valuable and practical tool not only for sportspeople and athletes [28,29] but also for individuals in their daily lives.

Based on the current knowledge, the effectiveness of ET applications in reducing neck pain and improving the ACROM has been demonstrated [30–32]. However, there is a lack of data on the immediate and short-term effects of a single bilateral and symmetrical ET application worn for three days by video terminal workers experiencing neck pain and restricted ACROM. The hypothesis of this research is that wearing the same ET application for three days can have beneficial effects on cervical pain and ACROM outcomes in video terminal workers. Therefore, the objective of this research was to assess the immediate and short-term impacts of a single bilateral and symmetrical ET application on self-perceived pain and multiplanar motion of the cervical spine.

2. Materials and Methods

2.1. Design and Participants

This study utilized a short-term longitudinal small-cohort design with repeated measures. The participants were recruited from employees of a call center located in the south of Italy. Voluntary participation was sought, and specific selection criteria were applied, including (1) individuals experiencing recurrent self-perceived musculoskeletal cervical pain (recurrence more than 1 episode per year), (2) individuals receiving no treatments for cervical pain before or during the research period, and (3) individuals working as video terminal operators.

A total of 60 participants were recruited and randomly assigned to either the study group (SG) or the control group (CG), ensuring a balanced distribution. However, 10 participants dropped out during the course of this study, resulting in a final sample size of 50 participants completing the experiment (Figure 1). The SG comprised 30 participants (*n:* 22-M, age 35.6 ± 4.4 years; body height 176.6 ± 6.6 cm; body mass 79.3 ± 9.1 kg and 8-F, age 34.8 ± 4.5 years; body height 163.5 ± 4.3 cm; body mass 57.6 ± 6.0 kg). The CG included 20 participants (*n:* 11-M, age 32.8 ± 3.6 years; body height 182.3 ± 7.5 cm; body mass 79.0 ± 9.2 kg and 9-F, 32.2 ± 4.5 years; body height 165.9 ± 6.8 cm; body mass 61.7 ± 10.4 kg). A power analysis was performed, indicating that sample sizes of 20 and 30 subjects per group, respectively, would provide 80% power, with a 5% error probability and an effect size of 0.55. Prior to the intervention phase, all participants were thoroughly informed of this study's purpose and provided voluntary consent. Privacy criteria were also strictly upheld. This study received approval from the Ovidius University of Constanta, Number 78, on 27 January 2023, in accordance with the principles outlined in the Helsinki Declaration.

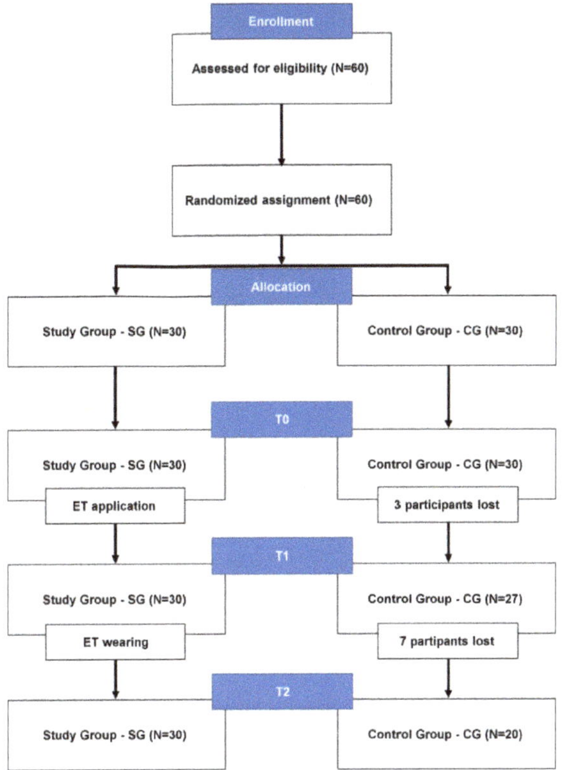

Figure 1. Participants' distribution into groups. In particular, for CG, (1) three participants lost immediately after T0 expressed discomfort when performing the ACROM test and were afraid that it might cause more pain. As a result, they chose to withdraw from this study. (2) Seven subjects were lost between T1 and T2 because they opted to undergo some form of treatment, and, therefore, they were excluded from this analysis.

2.2. Instrumentation

The assessment of multiplanar ACROM was conducted using an inertial sensor (Moover®, Sensor Medica, Guidonia-RM, Italy) positioned in the middle of the forehead and secured with an elastic band (Figure 2).

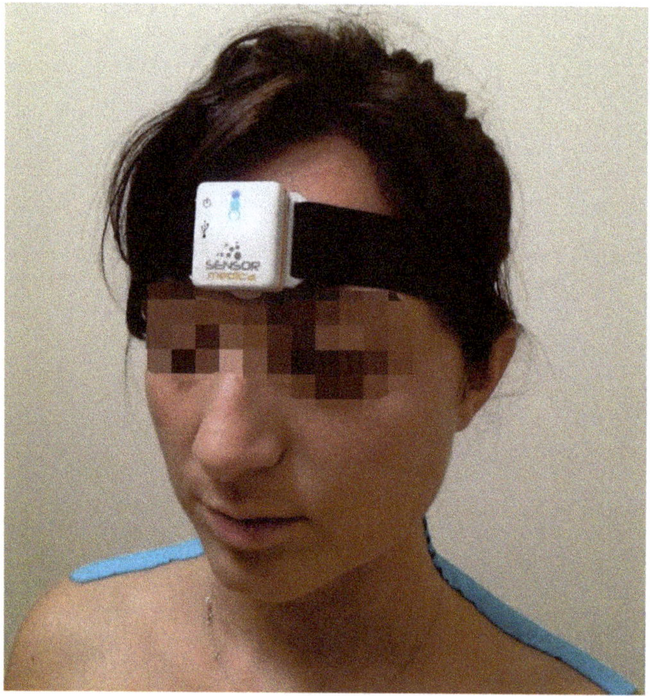

Figure 2. Placement of the inertial sensor on the forehead for measuring of the multiplanar ACROM.

To ensure the accuracy and validity of the inertial sensor, a preliminary comparison was made using a six three-dimensional camera optoelectronic system (SMART DX, BTS Bioengineering, Garbagnate Milanese, Italy), which served as the gold standard. A convenience sample of nineteen subjects participated in this trial, performing ACROM tests using the inertial sensor with a passive reflective marker attached to it. The data obtained from the Moover® sensor did not show any statistically significant differences compared to the 3D kinematic data. Further details can be found in Appendix A.

Self-perceived cervical pain was evaluated using the Numerical Rating Scale (NRS, 0–10), with 0 representing the absence of pain and 10 indicating the highest level of sustainable pain.

For the ET application, two personalized strips of Taping Elastico® (ATS, Arezzo, Italy) were used for each subject in the SG. The application procedures are described in the next section.

2.3. Procedure and Data Collection

The testing procedures were conducted in a dedicated room within the participants' working place, maintaining a mean temperature of 19 °C and a mean relative humidity of 52%. To minimize the potential influence of circadian effects, each subject underwent testing at the same time of day, as is customary in laboratory procedures of this nature [33]. This study consisted of three test sessions (T0–T1–T2), with a preliminary familiarization session conducted one week prior to the start of the protocol to provide instructions to the participants. T0 served as the baseline assessment, T1 served as the acute assessment taken 20 min after the application of the ET on the cervical area, and T2 served as the short-term assessment conducted after three days of wearing the ET application (Figure 3). T0 and T1 assessments were performed on the same day. Both SG and CG underwent the three test sessions in the same order. The CG received no intervention between T0 and T1, as well as between T1 and T2.

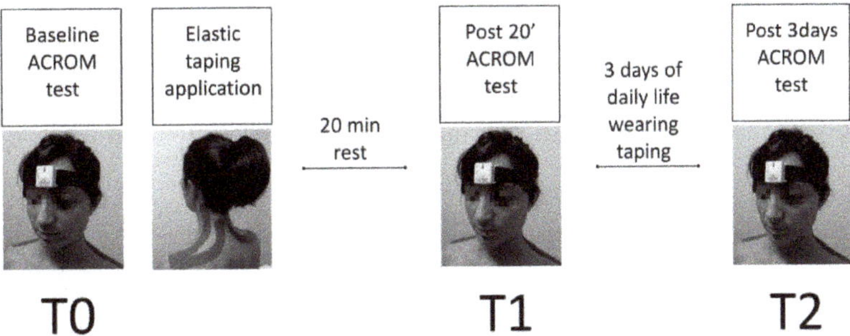

Figure 3. Graphical description of procedures and this study's timeline.

Each test session (T0–T1–T2) consisted of two evaluations performed in the following sequence: (1) the assessment of perceived cervical pain using a 0–10 Numerical Rating Scale (NRS) and (2) the measurement of ACROM using an inertial sensor.

The NRS was printed on white paper, and each subject marked an "X" on the score corresponding to their self-perceived pain in the cervical area.

The assessment of multiplanar ACROM involved measuring angular motions in three directions: right–left rotation on the transversal plane, right–left lateral inclination on the frontal plane, and flexion–extension on the sagittal plane (Figure 4). Each test included a total of 14 movements, with 7 repetitions performed for each direction. The maximum and minimum values were excluded, and the average value was calculated. During the tests, each subject was seated to stabilize the hips and the lumbar spine, while the entire trunk was leaned against a wall at shoulder height. Two flaps were placed on top of the wall to restrict shoulder motion. The tests were considered valid if the subject did not move the shoulders or trunk away from the wall. Only the cervical spine with the head were allowed to move. In case of an error, the test was repeated. The described procedure was also used in a previous study [11].

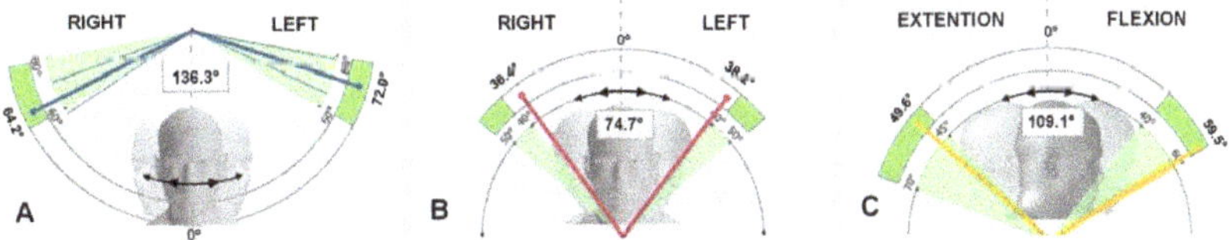

Figure 4. ACROM directions and anatomical planes tested using the inertial sensor: (**A**) right–left rotation on the transversal plane; (**B**) right–left lateral inclination on the frontal plane; (**C**) flexion–extension on the sagittal plane.

The ET application was administered by a skilled operator immediately after T0. The same operator (T.P.) applied the ET for all subjects in the study group (SG). The ET application followed the Taping Elastico® Method (ATS, Arezzo, Italy) and was applied symmetrically to both sides of the cervical area [34]. A strip of tape was cut into a "Y" shape, with the anchor applied to the skin at the level of the acromion and the two tails directed towards the base of the head. One tail followed the upper trapezius direction on the lateral portion of the neck, while the other was applied to the posterior portion (Figure 5). During the application, participants inclined their heads to the opposite side, stretching the skin. The ET was applied with zero tension, aiming to create convolutions when the head was in a neutral position. The ET application was bilateral and symmetrical.

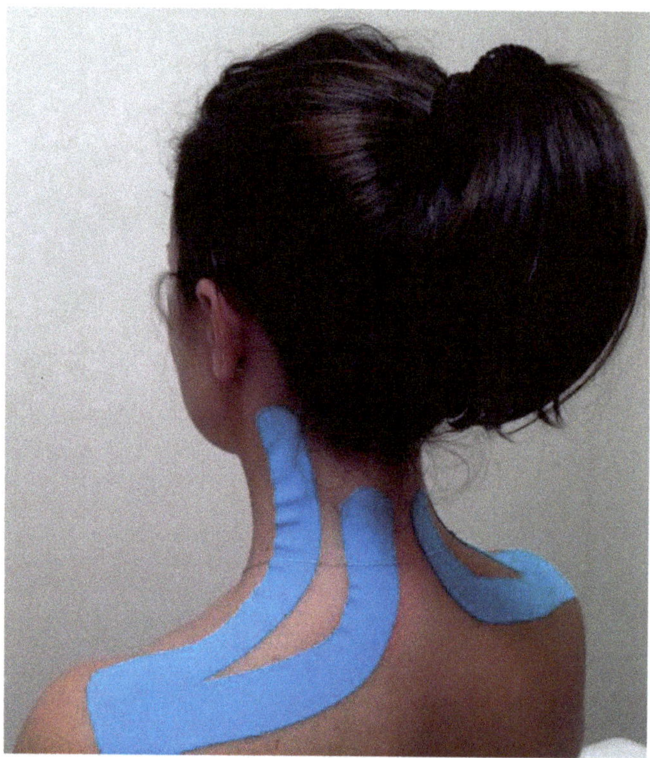

Figure 5. ET application following the Taping Elastico® Method. It was possible to distinguish the convolutions of the two tails, typical of a zero-tension application with the aim of reducing compression of the underlying tissues.

After the ET application, participants in the SG waited for 20 min while wearing the application before proceeding to the next test session (T1), which was in line with previous study [29] and the manufacturer's guidelines. The control group (CG) was observed at the same time between T0 and T1 without any intervention. Participants in the SG continued to wear the ET for three days during their daily activities. They were instructed to be cautious during dressing and washing to minimize the possibility of dislodging the ET. No instances of detachment were reported until T2. On the third day, one hour prior to the final test session (T2), the operator removed the ET.

2.4. Statistical Analysis

The normality of the data was assessed using the Shapiro–Wilk test. As the data followed a normal distribution, parametric tests were employed for the analysis. Differences at baseline were tested with a *t*-test for independent samples. A mixed ANOVA design (Time × Group) for repeated measures with Bonferroni correction was used to compare the post hoc effects (comparisons between T0 and T1, T1 and T2, and T0 and T2). Effect sizes (partial eta squared, η_p^2) were also calculated to facilitate the interpretation of the results, with values of 0.01, 0.06, and above 0.15 indicating small, medium, and large effect sizes, respectively [35,36]. The significance level was set at $p = 0.05$, and the statistical analysis was performed using SPSS (SPSS Inc., Chicago, IL, USA).

3. Results

The statistical analysis revealed no significant differences between SG and CG at T0 for any of the parameters measured. A Time $\times$ Group significant interaction was measured for all the ACROM directions, as well as for self-perceived pain, indicating that the changes in ROM and pain over time are not the same across the two groups (Table 1).

Table 1. Results of the mixed ANOVA Time $\times$ Group interaction.

Parameter	F	*p* Value
ACROM right rotation (°)	3.764	0.027
ACROM left rotation (°)	8.034	0.001
ACROM right lateral inclination (°)	10.400	0.000
ACROM left lateral inclination (°)	7.537	0.001
ACROM extension (°)	7.435	0.001
ACROM flexion (°)	5.708	0.005
Pain (A.U.)	113.111	0.000

Note: ACROM—active cervical range of motion.

The CG did not exhibit any significant differences in any parameter at any time point. However, the SG demonstrated a significant increase in all ACROM values at T2 compared to T0 (Table 2). Additionally, self-perceived pain in the cervical area was significantly lower at T2 compared to T0. Significant differences were also observed between T0 and T1, as well as between T1 and T2. Table 2 provides a detailed overview of the results.

Table 2. Results of the experiment for SG and CG.

	Parameter	T0	T1	T2	η_p^2	*p* Value
Study Group (SG)	ACROM right rotation (°)	66.0 ± 11.9	68.4 ± 9.3	74.2 ± 9.6 [#§]	0.432	0.000
	ACROM left rotation (°)	64.8 ± 7.7	70.4 ± 12.5 *	76.0 ± 11.1 [#§]	0.599	0.000
	ACROM right lateral inclination (°)	36.5 ± 7.9	38.0 ± 7.9	40.9 ± 5.2 [#§]	0.452	0.000
	ACROM left lateral inclination (°)	37.5 ± 6.9	40.6 ± 10.8	43.2 ± 8.0 [#§]	0.554	0.000
	ACROM extension (°)	47.0 ± 12.9	49.6 ± 11.3	55.1 ± 11.3 [#§]	0.360	0.000
	ACROM flexion (°)	55.0 ± 3.6	61.7 ± 12.8 *	62.9 ± 12.0 [#]	0.288	0.002
	Pain (A.U.)	7.5 ± 1.0	5.5 ± 1.4 *	1.4 ± 1.5 [#§]	0.945	0.000
Control Group (CG)	ACROM right rotation (°)	62.5 ± 9.1	63.8 ± 9.3	63.0 ± 8.7	0.017	0.569
	ACROM left rotation (°)	58.8 ± 7.0	60.9 ± 9.2	58.4 ± 7.5	0.012	0.630
	ACROM right lateral inclination (°)	36.0 ± 7.0	36.0 ± 6.7	34.1 ± 4.2	0.094	0.175
	ACROM left lateral inclination (°)	37.1 ± 7.1	36.4 ± 6.0	35.2 ± 6.6	0.009	0.675
	ACROM extension (°)	50.9 ± 11.6	52.9 ± 10.7	49.3 ± 10.8	0.049	0.337
	ACROM flexion (°)	56.5 ± 15.2	53.4 ± 14.1	54.3 ± 12.5	0.081	0.211
	Pain (a.u.)	6.9 ± 1.1	6.5 ± 1.7	7.1 ± 0.9	0.056	0.438

Note: ACROM—active cervical range of motion. T0—baseline test session. T1—test session after 20′ of wearing ET application. T2—test session after three days of wearing ET application. "*"—significant difference between T0 and T1. "#"—significant difference between T0 and T2. "§"—significant difference between T1 and T2.

The ET application resulted in increased ACROM in all directions after three days of wearing. However, significant increases were specifically observed in left rotation and flexion, even after just 20 min of wearing. Additionally, when comparing the measurements taken at T1 and T2, all ACROM directions, except for the extension phase, showed significant increases (Figure 6).

Furthermore, the self-perceived cervical pain exhibited a significant decrease after both 20 min and three days of wearing the ET application. Notably, there was also a significant difference in pain levels between T1 and T2 (Figure 7).

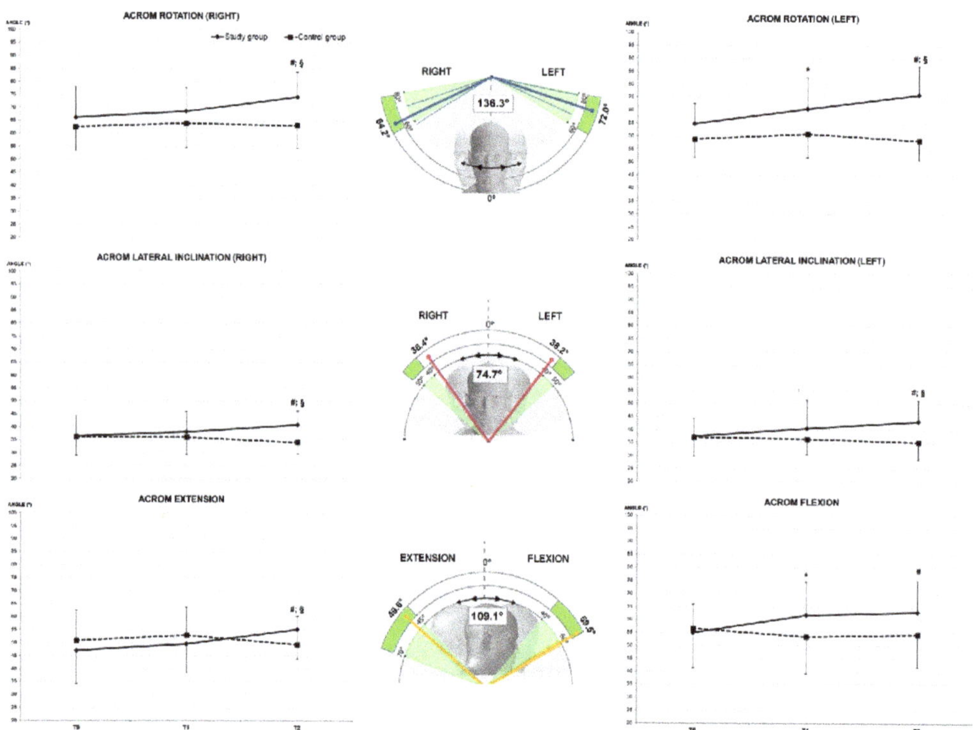

Figure 6. ACROM modifications over time in the six tested directions for SG (solid line) and CG (dotted line). "*"—significant difference between T0 and T1. "#"—significant difference between T0 and T2. "§"—significant difference between T1 and T2.

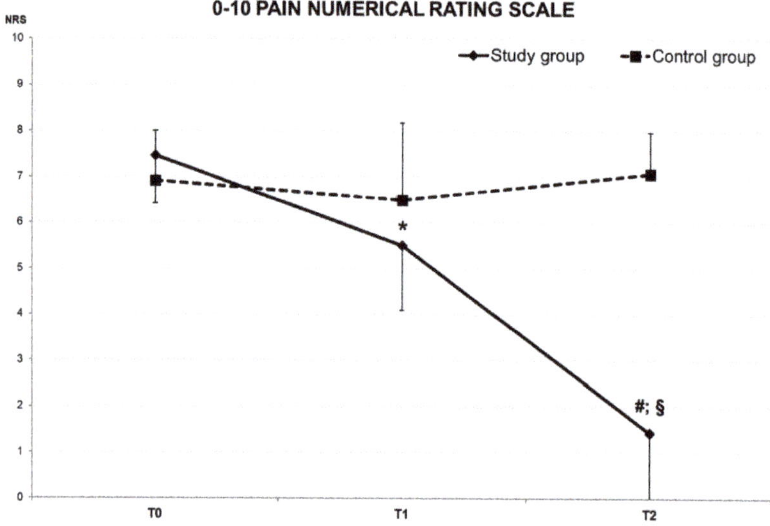

Figure 7. Numerical Rating Scale of self-perceived cervical pain modifications over time for SG (solid line) and CG (dotted line). "*"—Significant difference between T0 and T1. "#"—Significant difference between T0 and T2. "§"—Significant difference between T1 and T2.

Each group was analyzed by dividing them by gender, with the aim of understanding if the results of the entire sample could be influenced by gender differences. The behavior of male and female participants in both SG and CG was consistent with the behavior of the whole sample. SG showed significant differences between T0, T1, and T2 in both male and female individuals, while CG did not show any differences in any gender. Due to the paucity of female subjects in SG, the standard deviation was high, which affected the post hoc comparisons. Nevertheless, the ANOVA showed significant changes over time. Tables 3 and 4 summarize the statistical analyses for male and female individuals, respectively.

Table 3. Results of the experiment for male participants in SG and CG.

	Parameter	T0	T1	T2	η_p^2	*p* Value
Study Group (SG)	ACROM right rotation (°)	65.9 ± 11.0	68.2 ± 7.6	74.3 ± 7.4 [#§]	0.415	0.000
	ACROM left rotation (°)	64.7 ± 8.1	70.6 ± 9.9 *	75.9 ± 7.9 [#§]	0.680	0.001
	ACROM right lateral inclination (°)	35.9 ± 7.4	37.5 ± 7.4	40.7 ± 4.9 [#§]	0.536	0.000
	ACROM left lateral inclination (°)	36.0 ± 6.0	38.9 ± 7.6	42.5 ± 6.9 [#§]	0.626	0.000
	ACROM extension (°)	47.3 ± 13.1	51.7 ± 8.0	55.1 ± 9.4 [#]	0.327	0.004
	ACROM flexion (°)	55.6 ± 10.8	62.5 ± 11.3 *	63.2 ± 10.6	0.241	0.017
	Pain (A.U.)	7.5 ± 1.1	5.8 ± 1.3 *	1.5 ± 1.6 [#§]	0.938	0.000
Control Group (CG)	ACROM right rotation (°)	64.3 ± 9.0	64.4 ± 6.9	64.4 ± 6.7	0.002	0.892
	ACROM left rotation (°)	58.9 ± 7.2	62.7 ± 8.5	59.0 ± 6.9	0.000	0.948
	ACROM right lateral inclination (°)	38.2 ± 7.1	38.1 ± 5.7	35.6 ± 3.1	0.279	0.077
	ACROM left lateral inclination (°)	38.4 ± 7.4	38.5 ± 5.5	36.8 ± 6.9	0.141	0.229
	ACROM extension (°)	47.6 ± 13.6	50.0 ± 12.3	45.1 ± 11.2	0.079	0.376
	ACROM flexion (°)	57.8 ± 13.9	53.5 ± 13.0	56.0 ± 10.4	0.034	0.568
	Pain (a.u.)	4.3 ± 3.5	3.8 ± 3.3	4.6 ± 3.7	0.364	0.058

Note: ACROM—active cervical range of motion. T0—baseline test session. T1—test session after 20′ of wearing ET application. T2—test session after three days of wearing ET application. "*"—significant difference between T0 and T1. "#"—significant difference between T0 and T2. "§"—significant difference between T1 and T2.

Table 4. Results of the experiment for female participants in SG and CG.

	Parameter	T0	T1	T2	η_p^2	*p* Value
Study Group (SG)	ACROM right rotation (°)	66.5 ± 15.2	69.0 ± 13.6	73.7 ± 14.9	0.523	0.028
	ACROM left rotation (°)	65.1 ± 6.7	69.9 ± 18.8	76.3 ± 18.0	0.451	0.047
	ACROM right lateral inclination (°)	38.1 ± 9.5	39.4 ± 9.3	41.4 ± 6.3	0.256	0.165
	ACROM left lateral inclination (°)	41.6 ± 8.0	45.1 ± 16.7	45.3 ± 10.8	0.355	0.091
	ACROM extension (°)	46.2 ± 13.0	43.8 ± 16.8	55.1 ± 16.3 [§]	0.466	0.043
	ACROM flexion (°)	53.2 ± 12.2	59.5 ± 16.9	62.2 ± 16.2	0.484	0.037
	Pain (A.U.)	7.3 ± 0.7	4.8 ± 1.4 *	1.1 ± 1.5 [#§]	0.965	0.000
Control Group (CG)	ACROM right rotation (°)	60.4 ± 10.4	63.1 ± 12.0	61.1 ± 10.9	0.072	0.453
	ACROM left rotation (°)	58.7 ± 7.2	58.7 ± 10.0	57.7 ± 8.4	0.078	0.434
	ACROM right lateral inclination (°)	33.4 ± 6.2	33.3 ± 7.1	32.3 ± 4.9	0.129	0.308
	ACROM left lateral inclination (°)	35.5 ± 6.9	33.9 ± 5.9	33.2 ± 6.0	0.318	0.090
	ACROM extension (°)	54.8 ± 7.5	56.3 ± 7.7	54.4 ± 8.3	0.011	0.770
	ACROM flexion (°)	55.0 ± 17.3	53.3 ± 16.3	51.9 ± 15.0	0.185	0.215
	Pain (a.u.)	4.0 ± 3.9	4.1 ± 3.9	3.9 ± 3.5	0.053	0.594

Note: ACROM—active cervical range of motion. T0—baseline test session. T1—test session after 20′ of wearing ET application. T2—test session after three days of wearing ET application. "*"—significant difference between T0 and T1. "#"—significant difference between T0 and T2. "§"—significant difference between T1 and T2.

4. Discussion

The primary objective of this study was to evaluate the immediate and short-term impacts of the application of an ET to the cervical area on self-perceived pain and the multiplanar motion of the cervical spine. The main novelty of this investigation lies in examining the short-term effects of a bilateral and symmetrical ET application on ACROM.

This kind of approach can be considered a novelty because previous research did not use symmetrical applications. Additionally, investigating the effects of ET on pain is of great interest to healthcare professionals, given the chronic nature of cervical pain and its substantial societal costs [37–39]. In fact, cervical pain is recognized as one of the leading causes of global disability, placing it among the top five contributors [40].

Previous research conducted by Erdoğanoğlu et al. [30] with a similar study design demonstrated that wearing an ET application for 24 h resulted in a significant reduction in neck pain and improved ACROM. However, it is important to note that this previous study only included symptomatic individuals with cervical pain and lacked a control group; moreover, the observed effects were limited to a 24-h timeframe. In contrast, the present study also included symptomatic individuals but incorporated a control group and extended the measurement period by two days using a single ET application. This represents a significant advancement in the scientific evidence supporting the application of ET because it should be considered that the proposed methodology used in the current study is more similar to the everyday use of ET by individuals. Another study by Alahmari et al. [31] investigated the effects of ET application for more than three days, extending the period up to seven days, but it employed a different application technique and did not measure ACROM. Additionally, Ay et al. [32] demonstrated the effectiveness of five ET applications over a two-week period in terms of reducing neck pain and improving ACROM. While their study employed a similar ET application method to that used in our research, the ET shape, location, and unilateral application differed.

The findings of the present study align with previous literature (although different ET applications were used), supporting the immediate effects of ET on perceived pain and ACROM. Particularly noteworthy was the significant reduction in perceived pain after 20 min of ET application in the SG (7.5 ± 1.0 and 5.5 ± 1.4 at T0 and T1, respectively; $p < 0.000$). Moreover, in the same time span, there was a significant increase in ACROM for left rotation ($64.8 \pm 7.7°$ and $70.4 \pm 12.5°$ at T0 and T1, respectively, $p = 0.041$) and flexion ($55.0 \pm 3.6°$ and $61.7 \pm 12.8°$ at T0 and T1, respectively, $p = 0.007$). These immediate improvements in flexion and pain reduction confirm the positive effects of ET application. It is well recognized that individuals with cervical pain and disorders often experience limitations in flexion, making these findings particularly relevant [9–13].

From a practical and professional standpoint, the most notable finding of this study is undeniably the short-term effect observed after three days of using a single bilateral ET application. It is widely acknowledged that ET applications cannot be worn for extended periods due to factors such as personal hygiene practices, perspiration, and clothing changes. Consequently, it is common practice among healthcare professionals to remove and replace the ET application every three or four days [32,36]. The results of this study validate these procedural recommendations, demonstrating a substantial effect lasting for three days with a single ET application, without the need for removal and replacement.

For each direction of ACROM, a significant average improvement of 15% was observed between T0 and T2. Particularly noteworthy were the higher relative improvements observed in two specific directions: left rotation ($64.8 \pm 7.7°$ and $76.0 \pm 11.1°$ at T0 and T2, respectively, $p < 0.001$) and extension ($47.0 \pm 12.9°$ and $55.1 \pm 12.3°$ at T0 and T2, respectively, $p < 0.001$). These directions demonstrated a remarkable 17% increase in ACROM following three days of ET application. Furthermore, self-perceived pain exhibited a substantial average decrease of −81% between T0 and T2 (7.5 ± 1.0 and 1.4 ± 1.5, $p < 0.001$, respectively). It is relevant to highlight that the current results are gender independent, as both male and female participants showed significant improvements in ACROM and significant reductions in self-perceived pain. This aspect is highly relevant to public health because the absence of a gender effect enables professionals to apply the proposed method to a broad and diverse range of individuals without any gender-based restrictions.

One crucial aspect that deserves discussion in this paper is the different approach to the application of ET employed in this study compared to those of previous studies. While previous studies applied tension to the ET [30,31], with one exception that used a similar application method but a different placement and shape [32], in this study, the ET was applied without tension while stretching the skin during application (by inclining the head on the opposite side). This technique allowed the formation of skin convolutions when the head and neck were in a neutral position. It is very important to note that the two methods of application (with tension and without tension) differ significantly in terms of the pressure exerted on the skin. Although skin convolutions are thought to enhance local blood flow, the available data do not strongly support this claim [41–43]. While there are various studies of lower back pain that indicate no significant effects or differences between elastic taping applications with or without convolutions [44,45], there is a lack of similar data for the cervical area, except for the research conducted by Ay et al. [32]. They demonstrated that both convoluted and non-convoluted ET applications are effective in reducing neck pain and improving ACROM. Thus, the present study can be considered one of the first to use a bilateral and symmetrical ET application on the cervical area for three consecutive days, assessing its effects on self-perceived pain and ACROM. It is intriguing to hypothesize how two vastly different methods of ET (with and without tension) can yield similar results in terms of pain relief and cervical motion. A plausible explanation can be found in the existing literature, which suggests that the actual effect of ET can be attributed to both the direct contact itself and the pressure gradient generated between the taping and the surrounding skin area [46]. This idea is based on the findings of Pamuk et al. [47], who discuss the effects of ET on the underlying tissues, both at the immediate application site and in more distant regions. However, it should be noted that the latter explanation remains a hypothesis, and further research is required to elucidate this phenomenon. At present, the literature on ET continues to grapple with certain stigmas resulting from past inaccurate advertising that attributed false effects to taping. Therefore, studies like the present one are crucial for elucidating and enhancing the level of evidence regarding the use of ET for the treatment of musculoskeletal disorders.

Limitations

Like any scientific study, even this study has certain limitations that need to be acknowledged. One notable limitation is the inability to conduct additional test sessions following a three-day washout period of ET. Consequently, the findings of this paper are specific to the immediate and short-term effects of ET, and information regarding the residual effects after the removal is currently unavailable. This aspect is crucial because individuals with musculoskeletal disorders are often concerned about the duration of treatment's positive effects over time. Health professionals may find it valuable to comprehend the post-removal effects of ET and the duration of its benefits, enabling more precise intervention timing. Future research should aim to replicate our study protocol while incorporating a fourth and fifth test session to assess changes in self-perceived pain and multiplanar ACROM after ET removal. Since the restoration of muscle function is considered essential in the treatment of cervical spine disorders [15], it would be highly important to further investigate whether ET application can be regarded as beneficial for muscle function restoration.

The use of only one inertial sensor to detect cervical spine motion should be mentioned as a limitation. Although the inertial sensor showed results consistent with the 3D optoelectronic system and can be considered a valid tool for measuring cervical spine motion (see Appendix A), its use can introduce errors if the operator does not pay proper attention during the test. In fact, during this study, to reduce the margin of error, the shoulder position was fixed to avoid any compensatory motions. The authors strongly advise professionals to pay close attention to this aspect.

Another limitation of this study is the absence of a placebo group. Although previous results suggest the absence of a placebo effect for the cervical area [31,32] it would be interesting for future study designs to consider this critical aspect. For example, randomizing or having a crossover design would elicit some of this potential improvement.

In the end, although the current results can be considered gender independent because both male and female participants showed significant modifications induced via ET application, it is worth noting that the groups in this research are not gender-balanced. Specifically, the male-to-female ratio in this study and the control groups is 2.4 and 1.2, respectively. This imbalance is due to the participants lost during the preparation stage, as shown in Figure 1. It is possible that this aspect did not affect the results in any manner; however, it is fair to acknowledge its potential impacts. It would be interesting to replicate the experimental procedure with groups that have more balanced gender distributions.

5. Conclusions

In summary, the results of this study emphasize the effectiveness of wearing bilateral and symmetrical ET cervical applications for three days. To use ET is recommended for enhancing multiplanar ACROM and reducing self-reported cervical pain when needed, both in male and female subjects, especially among computer workers. This represents the primary novelty of this study. Specifically, applying ET without tension to create skin convolutions is a safe and cost-effective procedure for managing neck pain and improving the multiplanar motion of the cervical spine. These results can have immediate practical applications in the management of individuals' musculoskeletal health, especially among computer workers. Ultimately, a three-day ET application can be recommended for managing neck motion restrictions and pain in adult individuals. Nevertheless, further research is required to extend the application of these findings to other domains, such as sports.

Author Contributions: Conceptualization, L.R. and T.P.; methodology, L.R.; software, L.R. and P.B.; validation, A.L., J.P. and G.M.M.; formal analysis, L.R.; investigation, T.P. and A.R.; resources, A.L.; data curation, L.R.; writing—original draft preparation, L.R.; writing—review and editing, A.L., J.P. and G.M.M.; visualization, A.L., J.P. and G.M.M.; supervision, A.L., J.P. and G.M.M.; project administration, L.R. All authors have read and agreed to the published version of the manuscript.

Funding: This research received no external funding.

Institutional Review Board Statement: This study was conducted in accordance with the Declaration of Helsinki and approved by the Ethics Committee of Ovidius University of Constanta (n. 78, 27 January 2023).

Informed Consent Statement: Informed consent was obtained from all subjects involved in this study.

Data Availability Statement: The data that support the findings of this study are available from the corresponding author upon reasonable request.

Conflicts of Interest: Author Raggi A. was employed by the company FGP srl. The remaining authors declare that the research was conducted in the absence of any commercial or financial relationships that could be construed as a potential conflict of interest.

Appendix A

Appendix A contains the data referred to in the test session conducted to validate the Moover® inertial sensor in comparison to the 3D kinematics system. Table A1 presents a summary of the data obtained from the convenience sample. Table A2 provides the statistical analysis, while Figures A1–A6 depict the Bland–Altman graphs for each tested direction. These findings contribute to the assessment of the accuracy and validity of the inertial sensor in measuring multiplanar ACROM.

Table A1. Anthropometric of the sample (11M; 8 F).

Variables	Mean (SD)
Age (years)	36.4 (7.3)
Body Height (cm)	172.5 (7.8)
Body Mass (kg)	73.3 (19.3)

Table A2. Statistical comparison of the ACROM inertial sensor angular measures with 3D kinematics.

Parameter	Inertial Sensor Mean (SD)	3D Kinematics Mean (SD)	Z Score	*p* Value
Right rotation (°)	72.1 (11.7)	74.5 (11.6)	−0.818	0.414
Left rotation (°)	75.9 (8.5)	76.8 (9.2)	−0.336	0.737
Right lateral inclination (°)	44.0 (7.9)	46.2 (10.0)	−0.744	0.457
Left lateral inclination (°)	44.9 (7.2)	45.0 (6.4)	−0.248	0.804
Extension (°)	59.0 (10.0)	59.5 (10.9)	−0.044	0.965
Flexion (°)	63.2 (12.0)	63.6 (11.9)	−0.248	0.804

Due to the sample numerosity a Mann–Whitney test was used to compare the data collected via the two measurement systems. No significant differences were measured in ACROM between the two measuring systems in any tested direction.

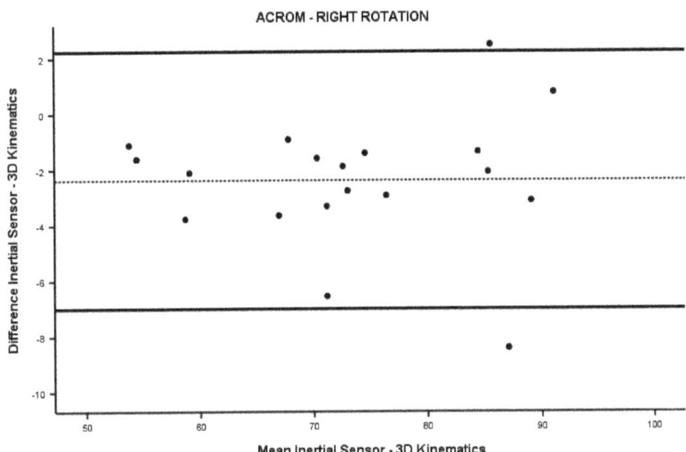

Figure A1. Bland–Altman graph for ACROM right rotation.

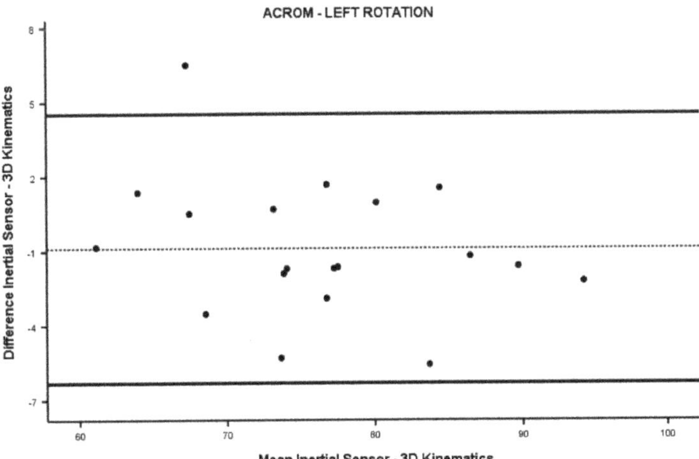

Figure A2. Bland–Altman graph for ACROM left rotation.

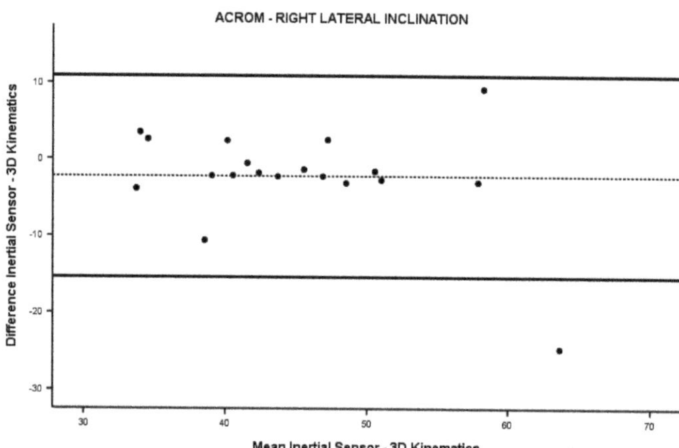

Figure A3. Bland–Altman graph for ACROM right lateral inclination.

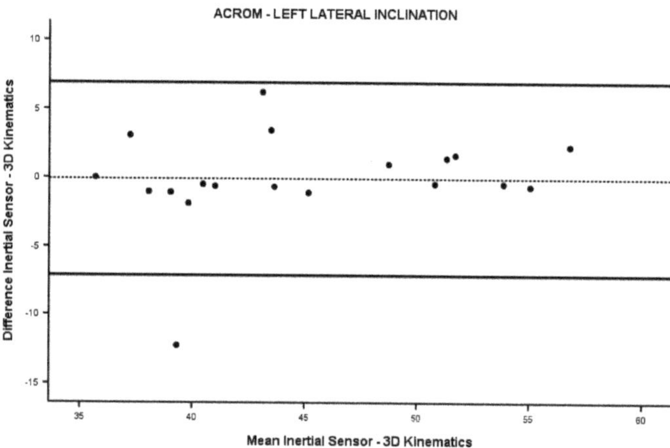

Figure A4. Bland–Altman graph for ACROM left lateral inclination.

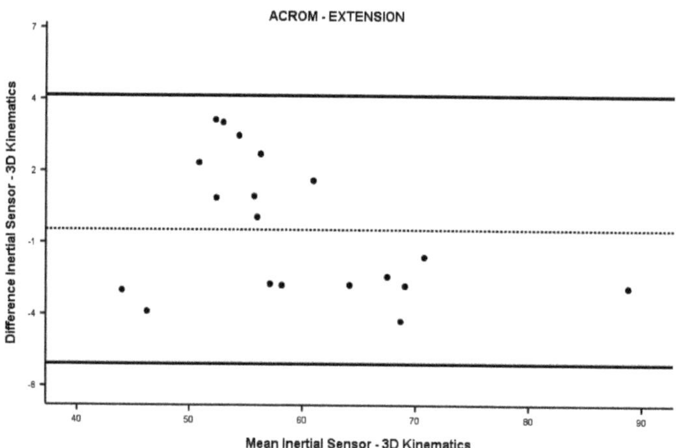

Figure A5. Bland–Altman graph for ACROM extension.

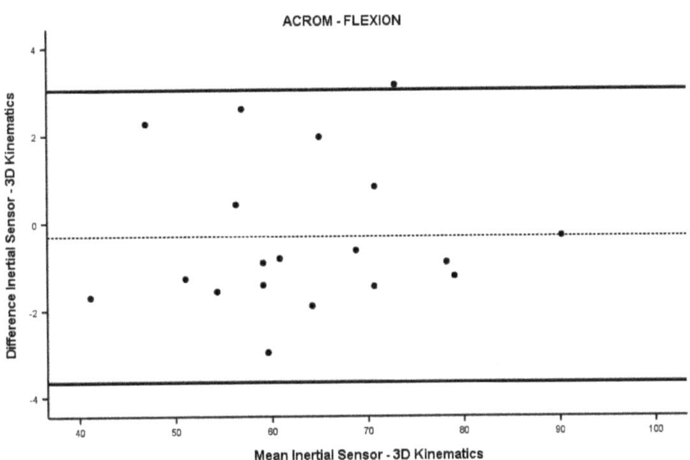

Figure A6. Bland–Altman graph for ACROM flexion.

References

1. Bogduk, N.; Mercer, S. Biomechanics of the Cervical Spine. I: Normal Kinematics. *Clin. Biomech.* **2000**, *15*, 633–648. [CrossRef]
2. Shugg, J.A.J.; Jackson, C.D.; Dickey, J.P. Cervical Spine Rotation and Range of Motion: Pilot Measurements during Driving. *Traffic Inj. Prev.* **2011**, *12*, 82–87. [CrossRef] [PubMed]
3. Thoomes-de Graaf, M.; Thoomes, E.; Fernández-de-las-Peñas, C.; Plaza-Manzano, G.; Cleland, J.A. Normative Values of Cervical Range of Motion for Both Children and Adults: A Systematic Review. *Musculoskelet. Sci. Pract.* **2020**, *49*, 102182. [CrossRef]
4. Dowdell, J.; Kim, J.; Overley, S.; Hecht, A. Biomechanics and Common Mechanisms of Injury of the Cervical Spine. *Handb. Clin. Neurol.* **2018**, *158*, 337–344. [CrossRef] [PubMed]
5. Lee, K.J.; Han, H.Y.; Cheon, S.H.; Park, S.H.; Yong, M.S. The Effect of Forward Head Posture on Muscle Activity during Neck Protraction and Retraction. *J. Phys. Ther. Sci.* **2015**, *27*, 977–979. [CrossRef] [PubMed]
6. Cheon, S.H.; Park, S.H. Changes in Neck and Upper Trunk Muscle Activities According to the Angle of Movement of the Neck in Subjects with Forward Head Posture. *J. Phys. Ther. Sci.* **2017**, *29*, 191–193. [CrossRef]
7. Gareiss, L.; Krumm, A.; Otte, A. On Biomechanics of the Cervical Spine When Using a Smartphone. *MMW Fortschr. Med.* **2020**, *162*, 10–14. [CrossRef]
8. Johnston, V.; Jull, G.; Souvlis, T.; Jimmieson, N.L. Neck Movement and Muscle Activity Characteristics in Female Office Workers with Neck Pain. *Spine* **2008**, *33*, 555–563. [CrossRef]
9. Hagen, K.B.; Harms-Ringdahl, K.; Enger, N.O.; Hedenstad, R.; Morten, H. Relationship between Subjective Neck Disorders and Cervical Spine Mobility and Motion-Related Pain in Male Machine Operators. *Spine* **1997**, *22*, 1501–1507. [CrossRef]
10. Jordan, A.; Mehlsen, J.; Ostergaard, K. A comparison of physical characteristics between patients seeking treatment for neck pain and age-matched healthy people. *J. Manip. Physiol. Ther.* **1997**, *20*, 468–475.
11. Woodhouse, A.; Vasseljen, O. Altered Motor Control Patterns in Whiplash and Chronic Neck Pain. *BMC Musculoskelet. Disord.* **2008**, *9*, 90. [CrossRef]
12. Stenneberg, M.S.; Rood, M.; de Bie, R.; Schmitt, M.A.; Cattrysse, E.; Scholten-Peeters, G.G. To What Degree Does Active Cervical Range of Motion Differ Between Patients With Neck Pain, Patients With Whiplash, and Those Without Neck Pain? A Systematic Review and Meta-Analysis. *Arch. Phys. Med. Rehabil.* **2017**, *98*, 1407–1434. [CrossRef]
13. Lee, H.; Nicholson, L.L.; Adams, R.D. Neck Muscle Endurance, Self-Report, and Range of Motion Data from Subjects with Treated and Untreated Neck Pain. *J. Manip. Physiol. Ther.* **2005**, *28*, 25–32. [CrossRef]
14. Williams, M.A.; McCarthy, C.J.; Chorti, A.; Cooke, M.W.; Gates, S. A Systematic Review of Reliability and Validity Studies of Methods for Measuring Active and Passive Cervical Range of Motion. *J. Manip. Physiol. Ther.* **2010**, *33*, 138–155. [CrossRef]
15. Jull, G.A.; O'Leary, S.P.; Falla, D.L. Clinical Assessment of the Deep Cervical Flexor Muscles: The Craniocervical Flexion Test. *J. Manip. Physiol. Ther.* **2008**, *31*, 525–533. [CrossRef]
16. Belli, G.; Toselli, S.; Mauro, M.; Maietta Latessa, P.; Russo, L. Relation between Photogrammetry and Spinal Mouse for Sagittal Imbalance Assessment in Adolescents with Thoracic Kyphosis. *J. Funct. Morphol. Kinesiol.* **2023**, *8*, 68. [CrossRef] [PubMed]
17. Russo, L.; Belli, G.; Di Blasio, A.; Lupu, E.; Larion, A.; Fischetti, F.; Montagnani, E.; Di Biase Arrivabene, P.; De Angelis, M. The Impact of Nordic Walking Pole Length on Gait Kinematic Parameters. *J. Funct. Morphol. Kinesiol.* **2023**, *8*, 50. [CrossRef] [PubMed]
18. Di Giminiani, R.; Di Lorenzo, D.; La Greca, S.; Russo, L.; Masedu, F.; Totaro, R.; Padua, E. Angle-Angle Diagrams in the Assessment of Locomotion in Persons with Multiple Sclerosis: A Preliminary Study. *Appl. Sci.* **2022**, *12*, 7223. [CrossRef]

19. Raya, R.; Garcia-Carmona, R.; Sanchez, C.; Urendes, E.; Ramirez, O.; Martin, A.; Otero, A. An Inexpensive and Easy to Use Cervical Range of Motion Measurement Solution Using Inertial Sensors. *Sensors* **2018**, *18*, 2582. [CrossRef]
20. Elizagaray-García, I.; Gil-Martínez, A.; Navarro-Fernández, G.; Navarro-Moreno, A.R.; Sánchez-De-toro-hernández, J.; Díaz-De-terán, J.; Lerma-Lara, S. Inter, Intra-Examiner Reliability and Validity of Inertial Sensors to Measure the Active Cervical Range of Motion in Patients with Primary Headache. *EXCLI J.* **2021**, *20*, 879–893. [CrossRef]
21. Russo, L.; Giustino, V.; Toscano, R.E.; Secolo, G.; Secolo, I.; Iovane, A.; Messina, G. Can Tongue Position and Cervical ROM Affect Postural Oscillations? A Pilot and Preliminary Study. *J. Hum. Sport Exerc.* **2020**, *15*, S840–S847. [CrossRef]
22. Theobald, P.S.; Jones, M.D.; Williams, J.M. Do Inertial Sensors Represent a Viable Method to Reliably Measure Cervical Spine Range of Motion? *Man Ther.* **2012**, *17*, 92–96. [CrossRef] [PubMed]
23. Wood, S.; Fryer, G.; Tan, L.L.F.; Cleary, C. Dry Cupping for Musculoskeletal Pain and Range of Motion: A Systematic Review and Meta-Analysis. *J. Bodyw. Mov. Ther.* **2020**, *24*, 503–518. [CrossRef] [PubMed]
24. Saavedra-Hernández, M.; Castro-Sánchez, A.M.; Arroyo-Morales, M.; Cleland, J.A.; Lara-Palomo, I.C.; Fernández-De-Las-Peñas, C. Short-Term Effects of Kinesio Taping versus Cervical Thrust Manipulation in Patients with Mechanical Neck Pain: A Randomized Clinical Trial. *J. Orthop. Sports Phys. Ther.* **2012**, *42*, 724–730. [CrossRef]
25. Xiong, J.; Zhang, Z.; Zhang, Z.; Ma, Y.; Li, Z.; Chen, Y.; Liu, Q.; Liao, W. Short-Term Effects of Kinesio Taping Combined with Cervical Muscles Multi-Angle Isometric Training in Patients with Cervical Spondylosis. *BMC Musculoskelet. Disord.* **2023**, *24*, 38. [CrossRef]
26. Šiško, P.K.; Videmšek, M.; Karpljuk, D. The Effect of a Corporate Chair Massage Program on Musculoskeletal Discomfort and Joint Range of Motion in Office Workers. *J. Altern. Complement. Med.* **2011**, *17*, 617–622. [CrossRef]
27. Tsauo, J.Y.; Lee, H.Y.; Hsu, J.H.; Chen, C.Y.; Chen, C.J. Physical Exercise and Health Education for Neck and Shoulder Complaints among Sedentary Workers. *J. Rehabil. Med.* **2004**, *36*, 253–257. [CrossRef]
28. Williams, S.; Whatman, C.; Hume, P.A.; Sheerin, K. Kinesio Taping in Treatment and Prevention of Sports Injuries: A Meta-Analysis of the Evidence for Its Effectiveness. *Sports Med.* **2012**, *42*, 153–164. [CrossRef]
29. Russo, L.; Bartolucci, P.; Ardigò, L.P.; Padulo, J.; Pausic, J.; Iacono, A. Dello An Exploratory Study on the Acute Effects of Proprioceptive Exercise and/or Neuromuscular Taping on Balance Performance. *Asian J. Sports Med.* **2018**, *9*, 63020. [CrossRef]
30. Erdoğanoğlu, Y.; Bayraklı, B. Short-Term Changes in Chronic Neck Pain After the Use of Elastic Adhesive Tape. *J. Chiropr. Med.* **2021**, *20*, 70–76. [CrossRef]
31. Alahmari, K.A.; Reddy, R.S.; Tedla, J.S.; Samuel, P.S.; Kakaraparthi, V.N.; Rengaramanujam, K.; Ahmed, I. The Effect of Kinesio Taping on Cervical Proprioception in Athletes with Mechanical Neck Pain-a Placebo-Controlled Trial. *BMC Musculoskelet. Disord.* **2020**, *21*, 648. [CrossRef]
32. Ay, S.; Konak, H.E.; Evcik, D.; Kibar, S. The Effectiveness of Kinesio Taping on Pain and Disability in Cervical Myofascial Pain Syndrome. *Rev. Bras. Reumatol.* **2017**, *57*, 93–99. [CrossRef]
33. Russo, L.; Montagnani, E.; Pietrantuono, D.; D'Angona, F.; Fratini, T.; Di Giminiani, R.; Palermi, S.; Ceccarini, F.; Migliaccio, G.M.; Lupu, E.; et al. Self-Myofascial Release of the Foot Plantar Surface: The Effects of a Single Exercise Session on the Posterior Muscular Chain Flexibility after One Hour. *Int. J. Environ. Res. Public Health* **2023**, *20*, 974. [CrossRef] [PubMed]
34. Russo, L.; Bartolucci, P. *Taping Elastico® Applicazioni in Chinesiologia*; ATS—Giacomo Catalani Editore: Arezzo, Italy, 2018.
35. Richardson, J.T.E. Eta squared and partial eta squared as measures of effect size in educational research. *Educ. Res. Rev.* **2011**, *6*, 135–147. [CrossRef]
36. Lakens, D. Calculating and reporting effect sizes to facilitate cumulative science: A practical primer for t-tests and ANOVAs. *Front. Psychol.* **2013**, *4*, 63. [CrossRef] [PubMed]
37. Korthals-de Bos, I.B.C.; Hoving, J.L.; Van Tulder, M.W.; Rutten-van Mölken, M.P.M.H.; Adèr, H.J.; De Vet, H.C.W.; Koes, B.W.; Vondeling, H.; Bouter, L.M. Cost Effectiveness of Physiotherapy, Manual Therapy, and General Practitioner Care for Neck Pain: Economic Evaluation alongside a Randomised Controlled Trial. *BMJ* **2003**, *326*, 911–914. [CrossRef] [PubMed]
38. Driessen, M.T.; Lin, C.W.C.; Van Tulder, M.W. Cost-Effectiveness of Conservative Treatments for Neck Pain: A Systematic Review on Economic Evaluations. *Eur. Spine J.* **2012**, *21*, 1441–1450. [CrossRef]
39. Aboagye, E.; Lilje, S.; Bengtsson, C.; Peterson, A.; Persson, U.; Skillgate, E. Manual Therapy versus Advice to Stay Active for Nonspecific Back and/or Neck Pain: A Cost-Effectiveness Analysis. *Chiropr. Man Therap.* **2022**, *30*, 27. [CrossRef]
40. Hoy, D.; March, L.; Woolf, A.; Blyth, F.; Brooks, P.; Smith, E.; Vos, T.; Jan, B.; Blore, J.; Murray, C.; et al. The Global Burden of Neck Pain: Estimates from the Global Burden of Disease 2010 Study. *Ann. Rheum. Dis.* **2014**, *73*, 1309–1315. [CrossRef]
41. Yang, J.M.; Lee, J.H. Is Kinesio Taping to Generate Skin Convolutions Effective for Increasing Local Blood Circulation? *Med. Sci. Monit.* **2018**, *24*, 288–293. [CrossRef]
42. Craighead, D.H.; Shank, S.W.; Volz, K.M.; Alexander, L.M. Kinesiology Tape Modestly Increases Skin Blood Flow Regardless of Tape Application Technique. *JPHR J. Perform. Health Res.* **2017**, *1*, 72–78. [CrossRef]
43. Banerjee, G.; Briggs, M.; Johnson, M.I. The Immediate Effects of Kinesiology Taping on Cutaneous Blood Flow in Healthy Humans under Resting Conditions: A Randomised Controlled Repeated-Measures Laboratory Study. *PLoS ONE* **2020**, *15*, e0229386. [CrossRef]
44. Parreira, P.D.C.S.; Costa, L.D.C.M.; Takahashi, R.; Junior, L.C.H.; da Luz Junior, M.A.; da Silva, T.M.; Costa, L.O.P. Kinesio Taping to Generate Skin Convolutions Is Not Better than Sham Taping for People with Chronic Non-Specific Low Back Pain: A Randomised Trial. *J. Physiother.* **2014**, *60*, 90–96. [CrossRef] [PubMed]

45. Araujo, A.C.; do Carmo Silva Parreira, P.; Junior, L.C.H.; da Silva, T.M.; da Luz Junior, M.A.; da Cunha Menezes Costa, L.; Pena Costa, L.O. Medium Term Effects of Kinesio Taping in Patients with Chronic Non-Specific Low Back Pain: A Randomized Controlled Trial. *Physiotherapy* **2018**, *104*, 149–151. [CrossRef] [PubMed]
46. Vercelli, S.; Colombo, C.; Tolosa, F.; Moriondo, A.; Bravini, E.; Ferriero, G.; Francesco, S. The Effects of Kinesio Taping on the Color Intensity of Superficial Skin Hematomas: A Pilot Study. *Phys. Ther. Sport* **2017**, *23*, 156–161. [CrossRef] [PubMed]
47. Pamuk, U.; Yucesoy, C.A. MRI Analyses Show That Kinesio Taping Affects Much More than Just the Targeted Superficial Tissues and Causes Heterogeneous Deformations within the Whole Limb. *J. Biomech.* **2015**, *48*, 4262–4270. [CrossRef]

Journal of
*Functional Morphology
and Kinesiology*

Article

Match Load Physical Demands in U-19 Professional Soccer Players Assessed by a Wearable Inertial Sensor

Guglielmo Pillitteri [1,2], Valerio Giustino [1], Marco Petrucci [2], Alessio Rossi [3,4], Ignazio Leale [1], Marianna Bellafiore [1], Ewan Thomas [1], Angelo Iovane [1], Antonio Palma [1,5] and Giuseppe Battaglia [1,5,*]

1   Sport and Exercise Sciences Research Unit, Department of Psychology, Educational Science and Human Movement, University of Palermo, 90144 Palermo, Italy
2   Palermo FC, 90046 Palermo, Italy
3   Department of Computer Science, University of Pisa, 56127 Pisa, Italy
4   National Research Council (CNR), Institute of Information Science and Technologies (ISTI), 56124 Pisa, Italy
5   Regional Sports School of CONI Sicilia, 90141 Palermo, Italy
*   Correspondence: giuseppe.battaglia@unipa.it

Abstract: Background: Wearable inertial sensors are poorly used in soccer to monitor external load (EL) indicators. However, these devices could be useful for improving sports performance and potentially reducing the risk of injury. The aim of this study was to investigate the EL indicators (i.e., cinematic, mechanical, and metabolic) differences between playing positions (i.e., central backs, external strikers, fullbacks, midfielders, and wide midfielder) during the first half time of four official matches (OMs). Methods: 13 young professional soccer players (Under-19; age: 18.5 ± 0.4 years; height: 177 ± 6 cm; weight: 67 ± 4.8 kg) were monitored through a wearable inertial sensor (TalentPlayers TPDev, firmware version 1.3) during the season 2021–2022. Participants' EL indicators were recorded during the first half time of four OMs. Results: significant differences were detected in all the EL indicators between playing positions except for two of them (i.e., distance traveled in the various metabolic power zones (<10 w) and the number of direction changes to the right >30° and with speed >2 m). Pairwise comparisons showed differences in EL indicators between playing positions. Conclusions: Young professional soccer players showed different loads and performances during OMs in relation to playing positions. Coaches should consider the different physical demands related to playing positions in order to design the most appropriate training program.

Keywords: soccer; sport performance; inertial sensor; inertial sensor device; inertial measurement unit; training load; external load; physical demand

Citation: Pillitteri, G.; Giustino, V.; Petrucci, M.; Rossi, A.; Leale, I.; Bellafiore, M.; Thomas, E.; Iovane, A.; Palma, A.; Battaglia, G. Match Load Physical Demands in U-19 Professional Soccer Players Assessed by a Wearable Inertial Sensor. *J. Funct. Morphol. Kinesiol.* **2023**, *8*, 22. https://doi.org/10.3390/jfmk8010022

Academic Editor: Abdul Rashid Aziz

Received: 30 November 2022
Revised: 23 January 2023
Accepted: 29 January 2023
Published: 7 February 2023

1. Introduction

Soccer is a situational team sport in which players are interconnected in a complex system characterized by technical–tactical components that are supported by physical and physiological factors [1]. In soccer, the physical performance consists of intermittent cyclic and acyclic activities characterized by aerobic and anaerobic demands [2,3]. The need to know the physical demands required during a soccer match is of fundamental importance for coaches and athletic trainers in order to properly plan the training program with the aim of increasing the training effect and reducing the risk of injury [4–7].

High-speed running, acceleration, and deceleration characteristics are considered determinant factors for physical performance and should be taken into consideration when designing the training program [8]. As a matter of fact, players usually perform 150–250 different actions and 1100 changes of direction during a match [9]. Moreover, players' physical activity characteristics change every 4–6 s based on the different player's positions on the pitch [9,10], resulting in a high rate of change in speed (i.e., acceleration). Studies have shown that professional players travel a total distance of between 10 and 13 km during a match [11]. Most of the total distance is covered at low intensities, whereas

22–24% is spent at intensities above 15 km/h, 8–9% above 20 km/h, and 2–3% above 25 km/h. Additionally, it has been found that players can perform between 600 and 650 accelerations during a match.

To improve performance while reducing the risk of injury, practitioners should assess match loads in relation to playing positions on the pitch to optimize training planning [12,13]. Specifically, the external load (EL) is usually described by the total distance, range of speed covered, accelerations, metabolic power [14], and other derived measures. Global position systems (GPS) technology has been largely used by practitioners to assess EL allowing for time-motion analysis in technical–tactical tasks [6,15–17]. However, in the last few years, several microelectromechanical systems (MEMS) have been developed and are available and these include triaxial accelerometers, triaxial gyroscopes, magnetometers, and pressure sensors. These devices, defined as inertial sensors devices (ISDs) or inertial measurement units (IMUs), can measure acceleration and angular velocity, among other parameters. ISD technology was developed specifically for the assessment in indoor sports where GPS devices cannot be used [18]. The literature suggests that this technology has not yet been fully used in professional soccer for EL monitoring.

It should be mentioned that physical demand is highly related to playing positions on the pitch due to the fact that roles have specific technical–tactical requests strictly related to different physical, physiological, energetic, and biomechanical components [2,13,19,20]. For example, the literature has reported that the total high-intensity distance is covered by central midfielders, wide defenders, and wide midfielders, whereas strikers and central defenders travel lower distances [21]. Moreover, wide defenders and wide midfielders perform the highest sprinting distance, whereas central midfielders and central backs cover the lowest in elite soccer players [21–25]. Moreover, significant differences were found between the various playing positions for all measures of EL in amateur soccer players [26]. Authors reported that central midfielders covered the longest distance during a match, which is in line with recent literature followed by the forwards, the full-backs, and the wide midfielders. As in adult professional soccer players, physical performance is affected by playing position also in youth elite players. In young male elite players (8–18 years), center backs covered the shortest high-intensity running and sprinting distances and wide midfielders the longest [27]. Additionally, in young soccer players (mean age 16.0 years), center defenders covered the shortest very high-speed ($\geq$19.8 km·h^{-1}) and sprint ($\geq$25.2 km·h^{-1}) distances, whereas wide players and center forwards covered the longest distances in these speed zones [28], which is in line with other studies [29].

It is worth noting that the GPS may have some signal issues due to adverse weather conditions or being used indoors compared to ISD. In fact, being that this technology is based on (inertial) movement, it allows for evaluating the EL considering the same parameters relating to distance, speed, and metabolic power with greater precision than GPS at 10 Hz [30]. Indeed, the better applicability of ISD technology is due to its small size, lower cost, and the possibility of using the device indoors, avoiding possible connection problems between GPS and satellite [30].

To the best of our knowledge, there are no studies that have investigated, through ISD technology, differences in EL indicators in young professional soccer players during official matches (OMs) considering the different playing positions on the pitch. We hypothesized that there may be differences in EL indicators in young professional soccer players during OMs depending on playing position. Hence, the aim of this study was to investigate any differences in EL indicators, specifically cinematic, mechanical, and metabolic indicators, measured through a wearable inertial sensor during the first half time of four OMs, considering the different playing positions in young professional soccer players (U19).

2. Materials and Methods

2.1. Study Design

In this cross-sectional study, young players from a professional Italian soccer club were monitored using a wearable inertial sensor during the first half time of four OMs during the season 2021–2022. Soccer players were categorized into five groups according to their playing position on the pitch as follows: central back (CB), external striker (ES), fullback (FB), midfielder (MD), and wide midfielder (WM).

2.2. Participants

Thirteen young professional soccer players (age: 18.5 ± 0.4 years; height: 177 ± 6 cm; weight: 67 ± 4.8 kg) competing in the Italian U19 Championship were included. Participants' playing positions were the following: CB ($n = 2$), ES ($n = 3$), FB ($n = 5$), MD ($n = 1$), and WM ($n = 2$). The following inclusion criteria were considered: (1) professional male soccer players belonging to the Under-19; (2) no injury in the previous six months. Based on the exclusion criteria, only goalkeepers were not eligible for the study.

All participants signed an informed consent form before taking part in the study. The study, which complies with the principles of the Declaration of Helsinki, was approved by the Bioethics Committee of the University of Palermo (n. 68/2021).

2.3. Procedures

Participants were monitored during four OMs in a regular pitch with a theoretical match density (m^2/player) referred to ~320 m^2 according to Riboli et al. [31,32]. OMs were played on a third-generation artificial pitch or natural grass. All participants performed a typical 25-min pre-match warm-up before each OM.

The ISD was started 5 min prior to the assessment. Data were collected through a wearable inertial sensor (TalentPlayers TPDev, firmware version 1.3) [30]. To avoid interunit errors, each participant was assigned the same ISD for each OM.

Among the available devices, TalentPlayers was chosen as it provides very similar data compared to traditional GPS systems (i.e., instantaneous speed and distance, change of directions, and metabolic data) and it is already used by various Italian soccer teams (https://talentplayers.com (accessed on 16 December 2016)). This ISD is a small wearable device integrating a six degrees of freedom MEMS inertial sensor, capable of providing real-time acceleration and rotation data along three orthogonal axes at a frequency of 100 Hz per channel. It is designed to be worn on the lower leg using an elastic band. The validity and reliability of the ISD have been previously reported [30].

All data were acquired by the TalentPlayer mobile app (software version 1.0.7) and uploaded to the TalentPlayers cloud.

EL indicators considered for this study were classified as cinematic, mechanical, and metabolic. All indicators assess the volume of OMs (except the metabolic power indicator) although some parameters represent intensity performance indicators. These are detailed in Table 1.

Table 1. Descriptions of the external load indicators.

Indicators	Type	Description (Unit of Measure)
TD	Cinematic/volume	Total distance covered (m)
MS *	Cinematic/intensity	Maximum speed reached (even for <1 s)
N°INTACC *	Mechanical/volume	Number of intense accelerations >2 m/s^2
N°INTDEC *	Mechanical/volume	Number of intense decelerations >2 m/s^2
TDA	Mechanical/volume	Distance traveled with positive acceleration (i.e., with speed increase) (m)
TDD	Mechanical/volume	Distance traveled with negative acceleration (i.e., with speed decrease) (m)
N°HSR *	Cinematic/volume	Number of high-intensity running at >20 km/h
WT	Cinematic/volume	Time spent in the various speed zones (<6 km/h) (s)
THSR *	Cinematic/volume	Time spent in the various speed zones (>20 km/h) (s)
WD	Cinematic/volume	Distance traveled in the various speed zones (<6 km/h) (m)

Table 1. *Cont.*

Indicators	Type	Description (Unit of Measure)
DHSR *	Cinematic/volume	Distance traveled in the various speed zones (>20 km/h) (m)
MP *	Metabolic/intensity	Metabolic Power ($w \cdot kg^{-1}$) was calculated by multiplying EC (in $J \cdot kg^{-1} \cdot m^{-1}$) by running speed (v; in $m \cdot s^{-1}$) at any given moment (i.e., every 0.2 s): P met = EC·v. In order to assess metabolic power, considering the energy expenditure and derived, the equation developed by di Prampero et al. [33] established on previously studies by Minetti et al. [34] and Osgnach et al. [14] was adopted. (Watt = w)
TLMP	Metabolic/volume	Time spent in various metabolic power zones (<10 w) (s)
THMP *	Metabolic/volume	Time spent in various metabolic power zones (20–35 w) (s)
TEMP *	Metabolic/volume	Time spent in various metabolic power zones (35–55 w) (s)
TMMP *	Metabolic/volume	Time spent in various metabolic power zones (>55 w) (s)
DLMP	Metabolic/volume	Distance traveled in the various metabolic power zones (<10 w) (m)
DHMP *	Metabolic/volume	Distance traveled in the various metabolic power zones (20–35 w) (m)
DEMP *	Metabolic/volume	Distance traveled in the various metabolic power zones (35–55 w) (m)
DMMP *	Metabolic/volume	Distance traveled in the various metabolic power zones (>55 w) (m)
N°CoDR *	Mechanical/volume	Number of direction changes to the right >30° and with speed >2 m/s
N°CoDL *	Mechanical/volume	Number of direction changes to the left >30° and with speed >2 m/s

Legend: TD, Total Distance; MS, Maximal Speed; N°INTACC, Number of Accelerations; N°INTDEC, Number of Decelerations; TDA, Total Distance Acceleration; TDD, Total Distance deceleration; N°HSR, Number of High-Speed Running; WT, Walking Time; THSR, Time High-Speed Running; WD, Walking Distance; DHSR, Distance High-Speed Running; MP, Metabolic Power; TLMP, Time Low Metabolic Power; THMP, Time High Metabolic Power; TEMP, Time Elevated Metabolic Power; TMMP, Time Max Metabolic Power; DLMP, Distance Low Metabolic Power; DHMP, Distance High Metabolic Power; DEMP, Distance Elevated Metabolic Power; DMMP, Distance Max Metabolic Power; N°CoDR, Number of Direction Changes to the Right; N°CoDL, Number of Direction Changes to the Left; * Intensity indicator.

2.4. Statistical Analysis

Normality distribution was calculated through the Shapiro–Wilk test. Means and standard deviations of all the EL indicators for OMs and for each playing position (CB, ES, FB, MD, WM) were provided.

A one-way analysis of variance (ANOVA) test on one factor (OM) was performed to detect differences for each EL indicator. The Tukey post hoc test was used for pairwise comparisons for each EL indicator between playing positions. Statistical significance was set at $p < 0.05$.

The Statistical Package jamovi (The jamovi project—jamovi Version 1.8.0.1) was used to perform data analysis. Graphs were created through Graph Pad Prism 8 (Version 8.0.2).

3. Results

Descriptive statistics of EL indicators (i.e., TD, MS, N°INTACC, N°INTDEC, TDA, TDD, N°HSR, WT, THSR, WD, DHSR, MP, TLMP, THMP, TEMP, TMMP, DLMP, DHMP, DEMP, DMMP, N°CoDR, N°CoDL) for each playing position (i.e., CB, ES, FB, MD, WM) are reported in Tables 2–5.

As reported in Table 6, results from the one-way ANOVA tests showed significant differences in all the EL indicators between playing positions (TD: $F_{(4,13,9)} = 16.59$, $p < 0.001$; MS: $F_{(4,11,8)} = 5.54$, $p = 0.009$; N°INTACC: $F_{(4,12,5)} = 6.22$, $p = 0.005$; N°INTDEC: $F_{(4,11,7)} = 3.43$, $p = 0.045$; TDA: $F_{(4,14,2)} = 9.99$, $p < 0.001$; TDD: $F_{(4,13,7)} = 23.29$, $p < 0.001$; N°HSR: $F_{(4,13,7)} = 6.94$, $p = 0.003$; WT: $F_{(4,13,1)} = 5.21$, $p = 0.01$; THSR: $F_{(4,14)} = 7.82$, $p = 0.002$; WD: $F_{(4,14)} = 4.76$, $p = 0.012$; DHSR: $F_{(4,14)} = 7.89$, $p = 0.002$; MP: $F_{(4,14,2)} = 8.83$, $p < 0.001$; TLMP: $F_{(4,12,6)} = 6.22$, $p = 0.005$; THMP: $F_{(4,14,1)} = 18.75$, $p < 0.001$; TEMP: $F_{(4,12,2)} = 4.1$, $p = 0.025$; TMMP: $F_{(4,13,1)} = 7.04$, $p = 0.003$; DHMP: $F_{(4,14,1)} = 15.51$, $p < 0.001$; DEMP: $F_{(4,12,3)} = 5.00$, $p = 0.013$; DMMP: $F_{(4,13,2)} = 7.6$, $p = 0.002$; N°CoDL:

$F(4,14) = 13.03$, $p < 0.001$) except for DLMP: $F(4,12,4) = 1.43$, $p = 0.28$ and N°CoDR: $F(4,13,9) = 2.93$, $p = 0.06$.

Table 2. Descriptive statistics of the external load indicators.

Cinematic						
TD (m)	MS (km/h)	N°HSR (Total)	WT (s)	THSR (s)	WD (m)	DHSR (m)
MEAN 5620	26.5	29.4	1451	54.7	1226	335
SD 537	2.58	10.3	179	24.5	155	152
MIN 4714	20.6	10	1106	13	987	76.5
MAX 6557	30.8	58	1780	143	1539	880

Mechanical						
N°INTACC (total)	N°INTDEC (total)	TDA (m)	TDD (m)	N°CoDR (total)		N°CoDL (total)
MEAN 25.5	24.9	3018	2582	143		139
SD 6.35	7.09	309	239	15.9		31.6
MIN 16	9	2466	2161	105		76
MAX 37	38	3569	3071	173		220

Metabolic								
MP (w)	TLMP (s)	THMP (s)	TEMP (s)	TMMP (s)	DLMP (m)	DHMP (m)	DEMP (m)	DMMP (m)
MEAN 10.7	1739	250	80.3	30.8	1883	998	399	179
SD 1.12	139	46.3	21.5	10.6	120	207	119	67.5
MIN 8.7	1502	172	33	14	1668	628	136	77.4
MAX 13	2004	337	133	56	2197	1398	711	329

Legend: TD, Total Distance; MS, Maximal Speed; N°INTACC, Number of Intense Accelerations; N°INTDEC, Number of Intense Decelerations; TDA, Total Distance Acceleration; TDD, Total Distance deceleration; N ° HSR, Number of High-Speed Running; WT, Walking Time; THSR, Time High-Speed Running; WD, Walking Distance; DHSR, Distance High-Speed Running; MP, Metabolic Power; TLMP, Time Low Metabolic Power; THMP, Time High Metabolic Power; TEMP, Time Elevated Metabolic Power; TMMP, Time Max Metabolic Power; DLMP, Distance Low Metabolic Power; DHMP, Distance High Metabolic Power; DEMP, Distance Elevated Metabolic Power; DMMP, Distance Max Metabolic Power; N°CoDR, Number of Change of Direction Right; N°CoDL, Number of Change of Direction Left; SD, standard deviation.

Table 3. Descriptive statistics of the external load indicators for each official match.

		Cinematic						
	MATCH	TD (m)	MS (km/h)	N°HSR (Total)	WT (s)	THSR (s)	WD (m)	DHSR (m)
MEAN	1	5708	26.8	29.8	1428	53.7	1248	328
	2	5673	25.7	31.6	1330	60.3	1151	369
	3	5456	26.8	26.6	1595	49.6	1325	303
	4	5625	26.8	29.3	1469	54.8	1187	335
SD	1	594	2.77	12.1	164	23.4	124	146
	2	625	1.97	12.4	163	35.0	109	216
	3	515	2.68	7.69	192	18.7	195	117
	4	444	3.14	8.86	96.8	19.3	153	119
MIN	1	4714	21.9	10	1177	13	1119	76.5
	2	4728	22.4	14	1106	20	987	120
	3	4936	23.3	14	1267	22	1079	131
	4	4740	20.6	11	1318	13	1026	77.4
MAX	1	6557	29.1	48	1705	81	1444	504
	2	6544	28.4	58	1575	143	1343	880
	3	6338	30.8	38	1780	82	1539	505
	4	6162	30.2	41	1571	78	1409	477

Table 3. *Cont.*

	MATCH	N°INTACC (total)	N°INTDEC (total)	TDA (m)	TDD (m)	N°CoDR (total)	N°CoDL (total)
		Mechanical					
MEAN	1	26.3	27.6	3097	2592	148	139
	2	24.8	23.1	3038	2616	138	146
	3	24.4	22.9	2903	2532	134	125
	4	26.4	25.9	3023	2581	150	145
SD	1	5.74	7.11	345	265	12.5	30.9
	2	6.89	7.94	356	275	18.3	33.3
	3	5.60	4.97	305	215	15.8	39.2
	4	7.93	7.97	229	228	12.8	21.8
MIN	1	18	18	2485	2206	124	101
	2	16	13	2466	2248	105	117
	3	19	15	2581	2280	108	76
	4	16	9	2557	2161	137	117
MAX	1	35	36	3569	2970	164	198
	2	35	38	3476	3071	162	220
	3	35	28	3476	2845	153	184
	4	37	35	3357	2904	173	188

	MATCH	MP (w)	TLMP (s)	THMP (s)	TEMP (s)	TMMP (s)	DLMP (m)	DHMP (m)	DEMP (m)	DMMP (m)
		Metabolic								
MEAN	1	10.9	1711	257	85.6	31.0	1896	1028	423	181
	2	11.1	1646	252	80.9	32.0	1867	1009	408	183
	3	10.1	1850	236	71.9	28.9	1907	938	356	168
	4	10.6	1766	254	82.0	31.1	1862	1012	406	182
SD	1	1.22	137	50.1	19.8	11.6	114	223	111	77.1
	2	1.26	122	46.6	29.1	10.9	152	219	163	69.0
	3	0.964	138	52.6	11.1	11.8	116	219	59.8	76.4
	4	0.884	80.7	41.1	23.0	9.46	106	190	126	57.3
MIN	1	8.88	1511	181	56	15	1738	690	254	79.5
	2	9.15	1502	172	48	21	1668	628	213	106
	3	9.05	1659	178	50	15	1743	688	242	81.8
	4	8.70	1641	187	33	14	1763	685	136	77.4
MAX	1	12.6	1958	321	107	48	2060	1319	559	289
	2	13.0	1857	337	133	56	2197	1398	711	329
	3	11.8	2004	327	87	52	2068	1324	450	312
	4	11.6	1864	327	101	44	2069	1330	511	261

Legend: TD, Total Distance; MS, Maximal Speed; N°INTACC, Number of Intense Accelerations; N°INTDEC, Number of Intense Decelerations; TDA, Total Distance Acceleration; TDD, Total Distance Deceleration; N° HSR, Number of High-Speed Running; WT, Walking Time; THSR, Time High-Speed Running; WD, Walking Distance; DHSR, Distance High-Speed Running; MP, Metabolic Power; TLMP, Time Low Metabolic Power; THMP, Time High Metabolic Power; TEMP, Time Elevated Metabolic Power; TMMP, Time Max Metabolic Power; DLMP, Distance Low Metabolic Power; DHMP, Distance High Metabolic Power; DEMP, Distance Elevated Metabolic Power; DMMP, Distance Max Metabolic Power; N°CoDR, Number of Change of Direction Right; N°CoDL, Number of Change of Direction Left; SD, standard deviation.

Table 4. External load indicator differences among playing positions (central back, external striker, full back, midfielder, wide midfielder).

	Midfielder (M ± SD)	Central Back (M ± SD)	External Striker (M ± SD)	Full Back (M ± SD)	Wide Midfielder (M ± SD)
			Cinematic		
TD (m)	5963 ± 74	5240 ± 340 [1,5]	5723 ± 438 [4]	5087 ± 344 [1,5]	6161 ± 316
MS (km/h)	25.7 ± 1.88	24.3 ± 2.66 [3,5]	28.4 ± 1.34 [4]	25 ± 2.61 [5]	28.2 ± 1.31
N°HSR	29.8 ± 3.5	18.7 ± 7.89 [3,5]	38.8 ± 8.88 [4]	22.9 ± 6.53 [5]	34.4 ± 7.03
WT (s)	1303 ± 78.5	1490 ± 60.5	1476 ± 205	1578 ± 187 [5]	1345 ± 151

Table 4. *Cont.*

	Midfielder (M ± SD)	Central Back (M ± SD)	External Striker (M ± SD)	Full Back (M ± SD)	Wide Midfielder (M ± SD)
			Cinematic		
THSR (s)	56 ± 5.72	29.7 ± 15.6 [3,5]	80.4 ± 26.3 [4]	39.9 ± 12.4	62.1 ± 13.5
WD (m)	1071 ± 53.9	1191 ± 161	1260 ± 137	1308 ± 186	1214 ± 129
DHSR (m)	342 ± 36.3	179 ± 95.2 [3,5]	494 ± 163 [4]	242 ± 77.3	381 ± 84.6
			Mechanical		
N°INTACC (total)	21.8 ± 4.35 [3]	22.5 ± 5.21 [3]	32.3 ± 3.99 [5]	25.4 ± 6.55	22.9 ± 5.3
N°INTDEC (total)	22.5 ± 6.86	21.7 ± 8.21	29.9 ± 3.91	21.9 ± 5	26.5 ± 8.64
TDA (m)	3136 ± 59.1 [3]	2818 ± 197 [5]	3067 ± 251 [4]	2732 ± 234 [5]	3348 ± 202
TDD (m)	2810 ± 64.4 [2,4]	2404 ± 159 [3,5]	2639 ± 206 [4]	2336 ± 117 [5]	2789 ± 125
N°CoDR (total)	139 ± 5.74	143 ± 8.99	146 ± 18.9	131 ± 19.1 [5]	153 ± 9.92
N°CoDL (total)	165 ± 6.08	152 ± 25.1	122 ± 25.1	149 ± 44.4	123 ± 15.4
			Metabolic		
MP (w)	11.1 ± 0.233	9.9 ± 0.792 [5]	11.1 ± 1.07 [4]	9.68 ± 0.771 [5]	11.7 ± 0.812
TLMP (s)	1625 ± 87.6 [4]	1793 ± 54	1754 ± 135	1856 ± 128 [5]	1626 ± 103
THMP (s)	286 ± 9.54 [2,4]	221 ± 32.7 [5]	245 ± 32.3 [5]	209 ± 23.7 [5]	301 ± 36.6
TEMP (s)	76.5 ± 17.7	67.2 ± 22.2 [3]	97.4 ± 16.6 [4]	66 ± 15.6	89.1 ± 19.2
TMMP (s)	25.3 ± 4.72 [3]	21.8 ± 7.25 [3,5]	42 ± 9.87 [4]	25 ± 7.98	34.9 ± 5.67
DLMP (m)	1790 ± 90.8	1888 ± 153	1894 ± 148	1934 ± 98.3	1863 ± 86.5
DHMP (m)	1139 ± 46.8 [2,4]	860 ± 151 [5]	999 ± 150 [5]	811 ± 110 [5]	1281 ± 164
DEMP (m)	382 ± 92	316 ± 111 [3]	502 ± 94.3 [4]	316 ± 82.9	450 ± 98.2
DMMP (m)	144 ± 31.4 [3]	120 ± 42.1 [3,5]	247 ± 59.5 [4]	140 ± 48.4	212 ± 43.9

Legend: TD, Total Distance; MS, Maximal Speed; N°INTACC, Number of Intense Accelerations; N°INTDEC, Number of Intense Decelerations; TDA, Total Distance Acceleration; TDD, Total Distance Deceleration; N°HSR, Number of High-Speed Running; WT, Walking Time; THSR, Time High-Speed Running; WD, Walking Distance; DHSR, Distance High-Speed Running; MP, Metabolic Power; TLMP, Time Low Metabolic Power; THMP, Time High Metabolic Power; TEMP, Time Elevated Metabolic Power; TMMP, Time Max Metabolic Power; DLMP, Distance Low Metabolic Power; DHMP, Distance High Metabolic Power; DEMP, Distance Elevated Metabolic Power; DMMP, Distance Max Metabolic Power; N°CoDR, Number of Change of Direction Right; N°CoDL, Number of Change of Direction Left. $p < 0.05$ for differences between playing positions ([1] difference with midfielder; [2] difference with central back; [3] difference with external striker; [4] difference with full back; [5] difference with wide midfielder); M, mean; SD, standard deviation.

Table 5. Highest and lowest external load indicator differences among playing positions (central back, external striker, full back, midfielder, wide midfielder).

	TD (m)	MS (km/h)	N°HSR (Total)	WT (s)	THSR (s)	WD (m)	DHSR (m)
				Cinematic			
Highest (M ± SD)	WM (6161 ± 316) MD (5963 ± 74)	ES (28.4 ± 1.34) WM (28.2 ± 1.31)	ES (38.8 ± 8.88) WM (34.4 ± 7.03)	FB (1578 ± 187) CB (1490 ± 60.5)	ES (80.4 ± 26.3) WM (62.1 ± 13.5)	FB (1308 ± 186) ES (1260 ± 137)	ES (494 ± 163) WM (381 ± 84.6)
Lowest (M ± SD)	FB (5087 ± 344) CB (5240 ± 340)	CB (24.3 ± 2.66) FB (25 ± 2.61)	CB (18.7 ± 7.89) FB (22.9 ± 6.53)	MD (1303 ± 78.5) WM (1345 ± 151)	CB (29.7 ± 15.6) FB (39.9 ± 12.4)	MD (1071 ± 53.9) CB (1191 ± 161)	CB (179 ± 95.2) FB (242 ± 77.3)
	N°INTACC (total)	N°INTDEC (total)	TDA (m)	TDD (m)	N°CoDR (total)	N°CoDL (total)	
				Mechanical			
Highest (M ± SD)	ES (32.3 ± 3.99) FB (25.4 ± 6.55)	ES (29.9 ± 3.91) WM (26.5 ± 8.64)	WM (3348 ± 202) MD (3136 ± 59.1)	MD (2810 ± 64.4) WM (2789 ± 125)	WM (153 ± 9.92) ES (146 ± 18.9)	MD (165 ± 6.08) CB (152 ± 25.1)	
Lowest (M ± SD)	MD (21.8 ± 4.35) CB (22.5 ± 5.21)	CB (21.7 ± 8.21) FB (21.9 ± 5)	FB (2732 ± 234) CB (2818 ± 197)	FB (2336 ± 117) CB (2404 ± 159)	FB (131 ± 19.1) MD (139 ± 5.74)	ES (122 ± 25.1) WM (123 ± 15.4)	

Table 5. *Cont.*

	MP (w)	TLMP (s)	THMP (s)	TEMP (s)	TMMP (s)	DLMP (m)	DHMP (m)	DEMP (m)	DMMP (m)
				Metabolic					
Highest (M ± SD)	WM (11.7 ± 0.812)	FB (1856 ± 128)	WM (301 ± 36.6)	ES (97.4 ± 16.6)	ES (42 ± 9.87)	FB (1934 ± 98.3)	WM (1281 ± 164)	ES (502 ± 94.3)	ES (247 ± 59.5)
	MD (11.1 ± 0.233)	CB (1793 ± 54)	MD (286 ± 9.54)	WM (89.1 ± 19.2)	WM (34.9 ± 5.67)	ES (1894 ± 148)	MD (1139 ± 46.8)	WM (450 ± 98.2)	WM (212 ± 43.9)
Lowest (M ± SD)	FB (9.68 ± 0.771)	MD (1625 ± 87.6)	FB (209 ± 23.7)	FB (66 ± 15.6)	CB (21.8 ± 7.25^3)	MD (1790 ± 90.8)	FB (811 ± 110)	FB (316 ± 82.9)	CB (120 ± 42.1)
	CB (9.9 ± 0.792)	WM (1626 ± 103)	CB (221 ± 32.7)	CB (67.2 ± 22.2)	FB (25 ± 7.98)	WM (1863 ± 86.5)	CB (860 ± 151)	CB (316 ± 111)	FB (140 ± 48.4)

Legend: TD, Total Distance; MS, Maximal Speed; N°INTACC, Number of Intense Accelerations; N°INTDEC, Number of Intense Decelerations; TDA, Total Distance Acceleration; TDD, Total Distance Deceleration; N°HSR, Number of High-Speed Running; WT, Walking Time; THSR, Time High-Speed Running; WD, Walking Distance; DHSR, Distance High-Speed Running; MP, Metabolic Power; TLMP, Time Low Metabolic Power; THMP, Time High Metabolic Power; TEMP, Time Elevated Metabolic Power; TMMP, Time Max Metabolic Power; DLMP, Distance Low Metabolic Power; DHMP, Distance High Metabolic Power; DEMP, Distance Elevated Metabolic Power; DMMP, Distance Max Metabolic Power; N°CoDR, Number of Change of Direction Right; N°CoDL, Number of Change of Direction Left; M, mean; SD, standard deviation.

Table 6. One-way ANOVA test results.

Indicators	F	df1	df2	p
TD	16.59	4	13.9	<0.001
MS	5.54	4	11.8	0.009
N°INTACC	6.22	4	12.5	0.005
N°INTDEC	3.43	4	11.7	0.045
TDA	9.99	4	14.2	<0.001
TDD	23.29	4	13.7	<0.001
N°HSR	6.94	4	13.7	0.003
WT	5.21	4	13.1	0.01
THSR	7.82	4	14	0.002
WD	4.76	4	14	0.012
DHSR	7.89	4	14	0.002
MP	8.83	4	14.2	<0.001
TLMP	6.22	4	12.6	0.005
THMP	18.75	4	14.1	<0.001
TEMP	4.1	4	12.2	0.025
TMMP	7.04	4	13.1	0.003
DLMP	1.43	4	12.4	0.28
DHMP	15.51	4	14.1	<0.001
DEMP	5	4	12.3	0.013
DMMP	7.6	4	13.2	0.002
N°CoDR	2.93	4	13.9	0.06
N°CoDL	13.03	4	14	<0.001

Legend: TD, Total Distance; MS, Maximal Speed; N°INTACC, Number of Intense Accelerations; N°INTDEC, Number of Intense Decelerations; TDA, Total Distance Acceleration; TDD, Total Distance deceleration; N°HSR, Number of High-Speed Running; WT, Walking Time; THSR, Time High-Speed Running; WD, Walking Distance; DHSR, Distance High-Speed Running; MP, Metabolic Power; TLMP, Time Low Metabolic Power; THMP, Time High Metabolic Power; TEMP, Time Elevated Metabolic Power; TMMP, Time Max Metabolic Power; DLMP, Distance Low Metabolic Power; DHMP, Distance High Metabolic Power; DEMP, Distance Elevated Metabolic Power; DMMP, Distance Max Metabolic Power; N°CoDR, Number of Change of Direction Right; N°CoDL, Number of Change of Direction Left. $p < 0.05$ for differences between playing positions.

Details of the Tukey post hoc analysis reporting the pairwise comparisons between playing positions are provided in Tables 6 and 7. The data show that all the EL indicators differ for each playing position except for N°INTDEC, WD, DLMP, and N°CoDL.

Table 7. Multiple comparison test results.

Cinematic Indicators	
TD	* significant difference between MD and CB ($p < 0.05$), between CM and FB ($p < 0.01$), between CB and WM ($p < 0.001$), and between FB and WM ($p < 0.001$).
MS	* significant difference between CB and ES ($p < 0.01$), between CB and WM ($p < 0.01$), between ES and FB ($p < 0.05$), and between FB and WM ($p < 0.05$).
N°HSR	* significant difference between CB and ES ($p < 0.001$), between CB and WM ($p < 0.01$), between ES and FB ($p < 0.01$), and between FB and WM ($p < 0.05$).
WT	* significant difference between FB and WM ($p < 0.05$).
THSR	* significant between CB and ES ($p < 0.001$), between CB and WM ($p < 0.05$), and between ES and FB ($p < 0.001$).
WD	no significant difference between playing positions.
DHSR	* significant between CB and ES ($p < 0.001$), between CB and WM ($p < 0.05$), and between ES and FB ($p < 0.001$).
Mechanical Indicators	
N°INTACC	* significant difference between MD and ES ($p < 0.05$), between CB and ES ($p < 0.05$), and between ES and WM ($p < 0.05$).
N°INTDEC	no significant difference between playing positions.
TDA	* significant difference between FB and MD ($p < 0.05$), between CB and MD ($p < 0.001$), between FB and ES ($p < 0.05$), and between WM and FB ($p < 0.001$)
TDD	* significant difference between MD and CB ($p < 0.01$), between MD and FB ($p < 0.001$), between CB and ES ($p < 0.05$), between CB and WM ($p < 0.001$), between ES and FB ($p < 0.01$), and between FB and WM ($p < 0.001$).
N°CoDR	* significant difference between FB and WM ($p < 0.05$).
N°CoDL	no significant difference between playing positions.
Metabolic Indicators	
MP	* significant difference between CB and WM ($p < 0.01$), between ES and FB ($p < 0.05$), and between FB and WM ($p < 0.001$).
TLMP	* significant difference between MD and FB ($p < 0.05$), and between FB and WM ($p < 0.001$).
THMP	* significant difference between MD and CB ($p < 0.05$), between MD and FB ($p < 0.01$), between CB and WM ($p < 0.001$), between ES and WM ($p < 0.01$), and between FB and WM ($p < 0.001$).
TEMP	* significant difference between CB and ES ($p < 0.05$), and between ES and FB ($p < 0.05$).
TMMP	* significant difference between ES and MD ($p < 0.01$), between ES and CB ($p < 0.001$), between CB and WM ($p < 0.05$), and between ES and FB ($p < 0.001$).
DLMP	no significant difference between playing positions.
DHMP	* significant difference between MD and CB ($p < 0.05$), between MD and FB ($p < 0.01$), between CB and WM ($p < 0.001$), between ES and WM ($p < 0.05$), and between FB and WM ($p < 0.001$).
DEMP	* significant difference between CB and ES ($p < 0.01$), and between ES and FB ($p < 0.001$).
DMMP	* significant difference between MD and ES ($p < 0.05$), between CB and ES ($p < 0.001$), between CB and WM ($p < 0.05$), between ES and FB ($p < 0.001$), and between FB and WM ($p < 0.05$).

Legend: TD, Total Distance; MS, Maximal Speed; N°INTACC, Number of Intense Accelerations; N°INTDEC, Number of Intense Decelerations; TDA, Total Distance Acceleration; TDD, Total Distance deceleration; N°HSR, Number of High-Speed Running; WT, Walking Time; THSR, Time High-Speed Running; WD, Walking Distance; DHSR, Distance High-Speed Running; MP, Metabolic Power; TLMP, Time Low Metabolic Power; THMP, Time High Metabolic Power; TEMP, Time Elevated Metabolic Power; TMMP, Time Max Metabolic Power; DLMP, Distance Low Metabolic Power; DHMP, Distance High Metabolic Power; DEMP, Distance Elevated Metabolic Power; DMMP, Distance Max Metabolic Power; N°CoDR, Number of Change of Direction Right; N°CoDL, Number of Change of Direction Left. * $p < 0.05$ for differences between playing positions.

Figures 1–3 show cinematic, mechanical, and metabolic indicators performed during OMs, respectively.

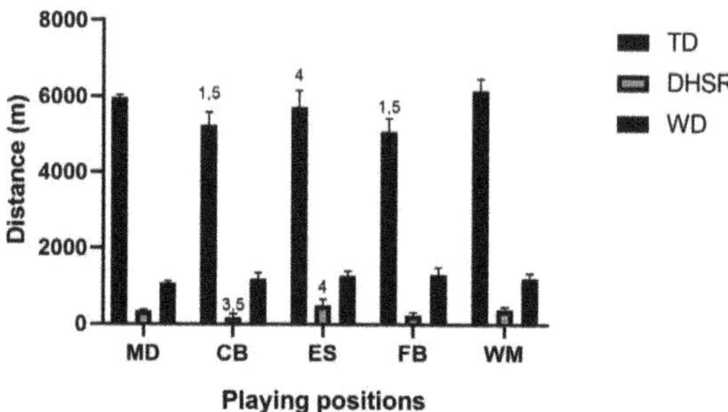

Figure 1. Playing position's cinematic indicators (TD, DHSR, WD) during the first half of official matches. Legend: TD, Total Distance; MS, WD, Walking Distance; DHSR, Distance High-Speed Running; MD, midfielder; CB, central back, ES, external striker, FB, fullback, WM, wide midfielder; $p < 0.05$ for differences between playing positions (1 difference with midfielder; 3 difference with external striker; 4 difference with full back; 5 difference with wide midfielder).

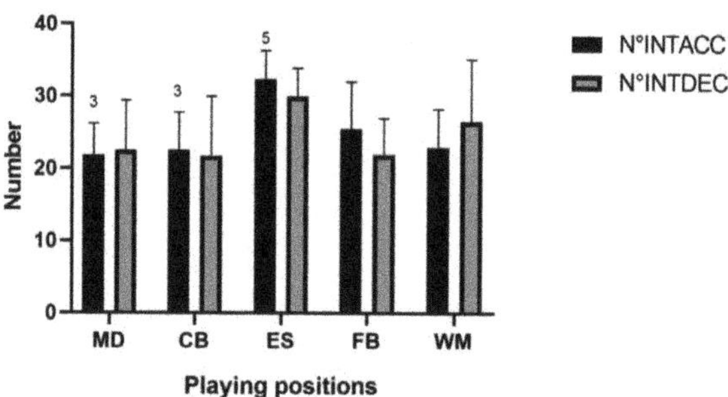

Figure 2. Playing position's mechanics indicators (N°INTACC, N°INTDEC) during the first half of official matches. Legend: N°INTACC, Number of Intense Accelerations; N°INTDEC, Number of Intense Decelerations; $p < 0.05$ for differences between playing positions (3 difference with external striker; 5 difference with wide midfielder).

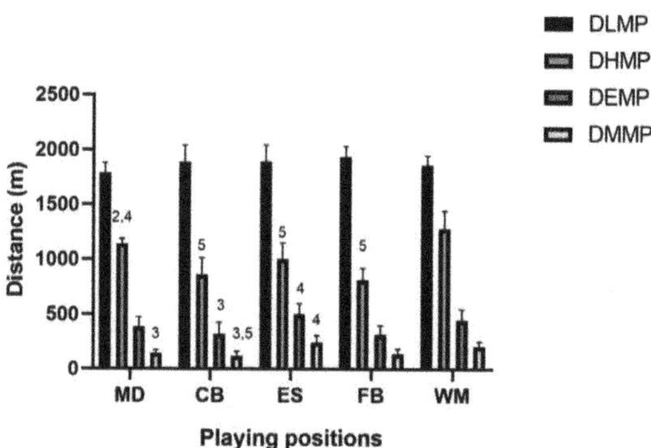

Figure 3. Playing position's metabolic indicators (DLMP, DHMP, DEMP, DMMP) during the first half of official matches. Legend: DLMP, Distance Low Metabolic Power; DHMP, Distance High Metabolic Power; DEMP, Distance Elevated Metabolic Power; DMMP, Distance Max Metabolic Power; $p < 0.05$ for differences between playing positions (2 difference with central back; 3 difference with external striker; 4 difference with full back; 5 difference with wide midfielder).

4. Discussion

The aim of this study was to investigate any differences in EL indicators during the first half time of four OMs between young professional soccer players based on their playing position. The EL indicators were assessed using a wearable inertial sensor device. This technology allows us to assess EL considering several indicators similar to GPS avoiding signal issues and connection problems. Moreover, Coutts et al. [35] have demonstrated that GPS devices are reliable in assessing total distance and peak speed during high-intensity intermittent exercise but are less reliable for high-intensity activities such as accelerations. In contrast, the ISD device tracks activity information using inertial technology that captures acceleration and rotation data in real time at a rate of 100 Hz per channel, resulting in more accuracy for detecting acceleration and high-intensity effort than GPS.

As we hypothesized, the main findings of our study showed significant differences in EL indicators between playing positions during OMs.

The scientific literature shows that physical demand in soccer players has been largely studied using GPS technology [17,36,37]. The main studies that assessed external load performance during matches found that physical performance is highly specific according to the role of the players. These results are in line with our findings, although we measured the EL indicators through an ISD. Indeed, considering cinematic, mechanical, and metabolic external load indicators, the ES, WM, and MD performed the highest level of physical performance during the first half time of the matches, whereas the CB and FB had the lowest level.

4.1. Cinematic External Load Indicators

In our study, significant differences in cinematic EL indicators between playing positions during the first half time of an OM were detected. Several previous studies have found differences in EL as distance, speed, and accelerations [38–40] between playing positions during both training and competition including both elite and amateur soccer players [26]. Moreover, considering physiological characteristics (i.e., HR and derived indices) [41,42] similar results have been found. The literature reports that the longest distance covered at high intensity has been achieved by the WM and FB [40–43]. In our study, we found that the WM and MD covered the highest TD during the first half time of an OM (6161 $\pm$ 316 and 5963 $\pm$ 74 m, respectively), whereas the CB (5240 $\pm$ 340 m) and FB (5087 $\pm$ 344 m) the lowest. Our results are in agreement with Ingebrigtsen et al., (2014) in which authors reported that, during the first half time of the match, WM traveled the highest TD and CB the lowest [44].

Concerning intensity indicators (i.e., N°HSR, THSR, and DHSR), ES and WM showed the highest results, whereas CB and FB showed the lowest. In line with these results, we found that FB and CB showed the highest amount of WT whereas MD and WM the lowest. However, there was no significant difference between the roles for the WD even though the FB and ES showed highest and the MD and CB the lowest values. Studies reported that CB showed the longest recoveries between consecutive high-intensity efforts [45] and spent the most time in low intensity efforts [39], in line with our results. Moreover, the MD spent less time in very low activity [46] and stand for much less time than other playing positions [47]. Indeed, MDs perform low to moderate-intensity activity more frequently showing shorter recovery bouts between high intensity efforts [46,47].

It is worth noting that when speed intensity increases, the MD exhibits low results. In fact, MDs play in very dense central spaces that limit the performance of intense actions such as high-speed running. However, when expressed as metabolic power, the central MD showed higher volume of high-intensity activity compared to attackers due to the accelerations [48].

The literature reports that FB perform more high intensity running than other playing positions. Specifically, Bangsbo et al., reported that MDs, FBs, and attackers covered a greater distance in high intensity running than the defenders [47–49]. Our results indicated that FB performed less intensity activity than other playing positions, showing similar performance to CB. Probably, in our sample, FB were required to have more defensive tactical function (i.e., 4:3:3 system) than offensive ones that requires a more intense effort during an OM. Indeed, some contextual factors such as tactics, game location, opponent quality, congested period, or match status could influence physical performance [50,51]. A study carried out by Altmann et al., (2021) [52] have demonstrated that FB (e.g., 4:4:2 or 4:2:3:1 system) displayed lower total and high-intensity distances compared to FB (e.g., 5:3:2 system), which is a new finding that emphasizes the need of differentiating between these two positions based on tactical system used. Moreover, WM, FB (in 3:5:2), and FB followed by forwards showed the greatest sprinting distance, whereas MD and CB showed shorter distances while sprinting. These findings are generally supported by previous literature [21,22,24,25].

Previous research indicates that FB, attackers, and MD players (both central and wide) covered the highest amounts of HSR and sprinting distance [25,53]. Dalen et al., (2016) reported that the FB and WM covered the highest whereas the CB the lowest HSR distance in an OM [54]. Ingebrigtsen et al., (2014) showed a greater HSR for the WM and FB and less for the CB in the first half of the match [44]. Additionally, authors reported the same results considering sprinting and distance [44]. In the same way, Dalen et al., (2016) reported highest sprinting distance for the FB and WM and less for the CB [54]. Moreover, Oliva Lozano et al., (2020) detected a sprinting distance greater for the WM and lower for the MD during an OM [8].

A high requirement for such running patterns in attacking players (e.g., WM) it may be necessary to cope with tactical demands related to overcoming defensive strategies to set up scoring situations. The MD plays in dense zones of the pitch that limit sprinting (i.e., speed >25.2 km/h) performance.

We also found that the players who reached the highest speed peaks were the ES and the WM, whereas CB and FB the lowest. These findings are in contrast with a previous study carried out by Rampinini et al., (2007) that reported a peak speed significantly higher for fullbacks than central backs during an OM [53]. Our results are in line with Oliva Lozano et al., (2020) [8] that indicate the highest peak speed for the WM and the lowest for the MD. The bigger space available for the WM compared to the CB and MD can explain these results.

4.2. Mechanical External Load Indicators

A determining factor in soccer performance is the acceleration profile [44,55]. Given the rate of change in speed performed by the players [56], the acceleration profile can be considered as a group of acceleration-based variables that requires a high neuromuscular physical demand [57]. Indeed, high-intensity accelerations and decelerations have a considerable impact on soccer players' mechanical load and can be counted as markers of muscle damage post-match [58]. Specifically, accelerations have a high metabolic cost [59], whereas decelerations increase the mechanical load [54].

Oliva Lozano et al., (2020) [8] have considered high intensity (i.e., >3 m/s^2) and low intensity (i.e., <3 m/s^2) accelerations. Authors reported that WM performed the highest and CB the lowest number of intense accelerations during an OM. However, different results have been found with a lower intensity threshold (i.e., low intensity accelerations). Indeed, the MD achieved the highest and WM the lowest number of accelerations. This result may be explained by the fact that density increases (reduced m^2 per player) as the ball is closer to the central zones of the pitch in match play resulting in a decrease in the intensity of play [32,60].

In our study, ES and FB performed a greater number of intense accelerations than other playing positions whereas the MD and CB are the lowest. The literature reported that in the first half of an OM, the highest number of accelerations have been showed by WM whereas the CB showed the lowest [54]. In this way, Ingebrigtsen et al., (2014) detected similar results. Authors reported that the WM and FB demonstrated a higher amount than the MD and CB [44].

Furthermore, our results confirm the findings of previous research which reported that the WM performed a higher number of intense accelerations than the CB [8].

Although the ES and WM showed the highest number of decelerations, there is no significant difference between the roles in our study.

Our results are in line with previous research where was reported that the WM performed higher number of intense decelerations than the CB [8]. Moreover, Oliva Lozano et al., (2020) found that the WM performed the highest number of intense decelerations and the CB the lowest during an OM, whereas the MD attained the highest and the WM the lowest number of low decelerations [8]. Moreover, Dalen et al., (2016) reported that the FB and WM performed a higher number of decelerations than the CB in the first half time of an OM [54].

We found that the WM and ES covered the highest TDA and FB and CB the lowest. This means that the FB performed shorter accelerations, similar to the CB, compared to the ES and WM.

In general, previous studies revealed that players in wide positions accelerated significantly more than central players [38,54]. Indeed, Oliva-Lozano et al., (2020) found that players covered longer acceleration distances in external positions (i.e., WM and FB) than the central MD and CB [8]. It is worth noting that Dalen et al., (2016) reported that the FB and WM covered the highest acceleration distance in the first half of the match whereas the CB was the lowest [54]. In addition, Abbott et al., (2018) found the highest intensity acceleration distances in wide positions (i.e., attackers and wide defenders) producing the highest distances due to the frequent tactical requirement of wide positions to reach high speeds [38].

In our study, whereas the MD and WM showed the highest decelerations distance, the FB and CB recorded the lowest. This result is in contrast with Dalen et al., (2016) in which a higher deceleration distance for CB and FB compared to WM in the first half of an OM was found [54]. However, there is a difference between the acceleration and deceleration thresholds used between the studies that do not allow us to compare them. Additionally, Oliva Lozano et al., (2020) reported that WM and forward covered the highest deceleration distance whereas the MD and CB were the lowest, accordingly to our results (except for the MD) [8]. However, different methods, MEMS technology, and thresholds used for classifying accelerations and decelerations make it difficult to draw conclusions about this difference.

In field-based intermittent sports such as soccer, it is proposed that the ability to execute rapid changes in direction is a critical factor in relation to match outcomes [61]. Therefore, the ability to make a quick change of direction is related to the ability to produce a large amount of force in a short time [62].

Research has shown that soccer players undertake approximately 700 direction changes of varying intensity during a match, and 600 of these changes in direction are 0–90° turns [2]. Around 50 of the direction changes are performed at maximal intensity during a match [2]. Approximately 700 direction changes per match were made by defenders, 500 by midfielders, and 600 by strikers. However, midfielders and strikers performed more turns of 270° to 360°. This could be due to specific tactical requirements such as playing position in midfield. The amount of 90° to 180° turns is relatively uniformly distributed with all positions performing roughly between 90 and 100 in official matches [2]. In our study, we considered a change of direction >30° and performed at a speed >2 m/s, classifying them in right and left changes of direction.

Regarding N°CoDR, we found that the difference appears to be significant only between the FB and WM. Specifically, the WM and ES performed the highest amount of CoDR, whereas the FB and MD the lowest. It is worth noting that, considering N°CoDL, there is no significant difference between the roles, even if the MD and CB performed the highest number whereas ES and WM were the lowest.

4.3. Metabolic External Load Indicators

Metabolic power, defined as the product of the energetic cost of acceleration running (EC, $J \cdot kg^{-1} \cdot m^{-1}$) and speed (v, $m \cdot s^{-1}$), has been considered for EL monitoring as it considers both speed and acceleration factors [14].

Studies that assessed the performance considering only the speed category have underestimated the amount of high-intensity activity performed by players. Indeed, when expressed as metabolic power, Gaudino et al., (2013) indicated that the MD showed a higher volume of high-intensity activity compared to attackers [48]. Our study confirms previous findings showing that midfielders showed higher metabolic power during an OM compared to other playing positions.

Specifically, our results show that the playing positions that recorded the highest MP average values were the WM, followed by the MD and ES, whereas the FB and CB were the lowest. In line with our results, Manzi et al., (2022) [63] reported that central backs covered less high-metabolic power distance and performed lower power events than players in the other playing positions, a finding likely related to the tactical role. Furthermore, midfielders covered both a substantial distance at high metabolic power and the largest number of power events compared to other playing positions.

This result could be supported by the fact that midfielders play in very dense central spaces of midfield that limit the performance of intense actions such as high-speed running and possibly increase the number of accelerations. It is worth noting that metabolic power increases because of speed or accelerations [14], consequently, since the MDs do not reach high distances at high intensity, their metabolic power increases more due to the numerous accelerations and decelerations. Furthermore, these results are in line with the results of a previous study showing that midfielders spend most of their time in medium and high-intensity activity during a match [39]. We also found that the MD and WM performed the highest distance per minute, whereas the CB and FB showed the lowest values. This result could be explained by the low recovery time and distance (i.e., TLMP and DLMP) among the actions detected in midfielders.

In addition, supporting our results, a study reported that midfielders had greater power recovery than central backs and forwards and lower recovery time after power events than central back, full back, and forward players [63]. Our study provided a detailed analysis of metabolic power as different metabolic intensity thresholds were reported for each playing position.

In line with previous findings, we found that the CB showed the lowest high, elevated, and maximum MP (expressed as both time and distance), whereas the MD and WM showed the highest THMP. However, considering the "intense" power thresholds (i.e., TEMP, TMMP, DEMP, and DMMP), the highest results were found in the ES and WM. In order to reach elevated and maximal metabolic power, players have to reach high speeds. This is possible by having space available as in the case of wide players such as the ES and WM. Indeed, Di Pampero et al., (2005) showed that the peak power output, of about 100 W/kg, is attained after about 0.5 s and that the average power over the first 4 s is on the order of 65 W/kg during 30 m running [33]. This means that players have to do a large number of intense long straight intense runs to achieve high and maximum metabolic power.

4.4. Strengths and Limitations of the Study

In this study, we considered cinematic, mechanical, and metabolic EL indicators that provide a complete overview of match performance and permit the evaluation of differences between playing positions.

Furthermore, it is worth mentioning that a different tactical requirement, the score of the match, the quality of the opponent, and the fact that the present study is based on only four official matches considering only thirteen participants represent the limitations of this study.

4.5. Practical Implications

In this study, we used a wearable inertial sensor (ISD technology) that provides several EL indicators useful to assess training and match performance in soccer. Compared to GPS, ISD can also be used for indoor sports without the need of coupling with external signals.

Moreover, data sampling takes place differently. ISD measures movement in real time through limb swing, whereas GPS uses the Doppler effect of satellite signals which could increase the distortion of related signals.

In summary, these devices can be useful for practitioners in assessing soccer players' EL during training or competitions in order to prevent injuries and improve sports performance.

5. Conclusions

The present study provides a useful and novel insight into sprint and acceleration profiles of young Italian professional soccer players during OMs showing a difference in EL indicators between playing positions. In particular, considering cinematic, mechanical, and metabolic EL indicators, the ES, WM, and MD performed the highest level of physical performance during a match, whereas the CB and FB performed the lowest.

However, there is high variability in physical performance during a match between players of the same position [24] (e.g., FB) and a possible explanation for this observation could be that the match performance depends not only on playing position but also to some extent on individual players themselves [52].

This study reveals some new findings concerning the physical demand for each playing position during OMs. To design role-specific soccer training, coaches need a clear view of how different players and positions meet physical requirements. The principle of specificity (SAID) suggests that the training process should be designed to emphasize certain physical components to cope with match demands. Wide players exhibit greater sprinting distances than central players. Finally, as with sprinting, wide players seem to perform more accelerations than central players [12].

Author Contributions: Conceptualization, G.P. and G.B.; methodology, V.G.; software, M.P. and A.R.; validation, A.I., M.B. and A.P.; formal analysis, A.R.; investigation, M.P. and I.L.; resources, G.B.; data curation, E.T.; writing—original draft preparation, G.P.; writing—review and editing, V.G.; visualization, A.R.; supervision, G.B.; project administration, A.I.; funding acquisition, A.P. All authors have read and agreed to the published version of the manuscript.

Funding: This research received no external funding.

Institutional Review Board Statement: The study was conducted according to the Declaration of Helsinki and approved by the Bioethics Committee of the University of Palermo (n. 68/2021).

Informed Consent Statement: Informed consent was obtained from all participants involved in the study.

Data Availability Statement: The datasets generated during and/or analyzed during the current study are available from the corresponding author on reasonable request.

Acknowledgments: The authors acknowledge all the players involved in the study. Furthermore, the authors thank the soccer managers and the FC Palermo soccer team.

Conflicts of Interest: The authors declare no conflict of interest.

References

1. Mota, T.; Silva, R.; Clemente, F. Holistic soccer profile by position: A theoretical framework. *Hum. Mov.* **2021**, *24*, 1–17. [CrossRef]
2. Bloomfield, J.; Polman, R.; O'Donoghue, P. Physical Demands of Different Positions in FA Premier League Soccer. *J. Sports Sci. Med.* **2007**, *6*, 63–70. [PubMed]
3. Dolci, F.; Hart, N.H.; Kilding, A.E.; Chivers, P.; Piggott, B.; Spiteri, T. Physical and Energetic Demand of Soccer: A Brief Review. *Strength Cond. J.* **2020**, *42*, 70–77. [CrossRef]
4. Jeffries, A.C.; Marcora, S.M.; Coutts, A.J.; Wallace, L.; McCall, A.; Impellizzeri, F.M. Development of a Revised Conceptual Framework of Physical Training for Use in Research and Practice. *Sports Med.* **2021**, *52*, 709–724. [CrossRef]

5. Buchheit, M.; Laursen, P.B. High-intensity interval training, solutions to the programming puzzle. Part II: Anaerobic energy, neuromuscular load and practical applications. *Sports Med.* **2013**, *43*, 927–954. [CrossRef]
6. Rossi, A.; Pappalardo, L.; Cintia, P.; Iaia, F.M.; Fernàndez, J.; Medina, D. Effective injury forecasting in soccer with GPS training data and machine learning. *PLoS ONE* **2018**, *13*, e0201264. [CrossRef]
7. Gabbett, T.J.; Nassis, G.P.; Oetter, E.; Pretorius, J.; Johnston, N.; Medina, D.; Rodas, G.; Myslinski, T.; Howells, D.; Beard, A.; et al. The athlete monitoring cycle: A practical guide to interpreting and applying training monitoring data. *Br. J. Sports Med.* **2017**, *51*, 1451–1452. [CrossRef]
8. Oliva-Lozano, J.M.; Fortes, V.; Krustrup, P.; Muyor, J.M. Acceleration and sprint profiles of professional male football players in relation to playing position. *PLoS ONE* **2020**, *15*, e0236959. [CrossRef]
9. Krustrup, P.; Mohr, M.; Ellingsgaard, H.; Bangsbo, J. Physical Demands during an Elite Female Soccer Game: Importance of Training Status. *Med. Sci. Sports Exerc.* **2005**, *37*, 1242–1248. [CrossRef]
10. Chmura, P.; Konefał, M.; Chmura, J.; Kowalczuk, E.; Zając, T.; Rokita, A.; Andrzejewski, M. Match outcome and running performance in different intensity ranges among elite soccer players. *Biol. Sport* **2018**, *35*, 197–203. [CrossRef]
11. Stølen, T.; Chamari, K.; Castagna, C.; Wisløff, U. Physiology of soccer: An update. *Sports Med.* **2005**, *35*, 501–536. [CrossRef]
12. Baptista, I.; Johansen, D.; Figueiredo, P.; Rebelo, A.; Pettersen, S.A. Positional Differences in Peak- and Accumulated- Training Load Relative to Match Load in Elite Football. *Sports* **2019**, *8*, 1. [CrossRef]
13. Brito, Â.; Roriz, P.; Duarte, R.; Garganta, J. Match-running performance of young soccer players in different game formats. *Int. J. Perform. Anal. Sport* **2018**, *18*, 410–422. [CrossRef]
14. Osgnach, C.; Poser, S.; Bernardini, R.; Rinaldo, R.; Di Prampero, P.E. Energy cost and metabolic power in elite soccer: A new match analysis approach. *Med. Sci. Sport. Exerc.* **2010**, *42*, 170–178. [CrossRef]
15. Ehrmann, F.; Duncan, C.S.; Sindhusake, D.; Franzsen, W.N.; Greene, D.A. GPS and Injury Prevention in Professional Soccer. *J. Strength Cond. Res.* **2015**, *30*, 360–367. [CrossRef]
16. Van Eetvelde, H.; Mendonça, L.D.; Ley, C.; Seil, R.; Tischer, T. Machine learning methods in sport injury prediction and prevention: A systematic review. *J. Exp. Orthop.* **2021**, *8*, 27. [CrossRef]
17. Varley, M.; Fairweather, I.H.; Aughey, R.J. Validity and reliability of GPS for measuring instantaneous velocity during acceleration, deceleration, and constant motion. *J. Sports Sci.* **2012**, *30*, 121–127. [CrossRef]
18. Falbriard, M.; Meyer, F.; Mariani, B.; Millet, G.P.; Aminian, K. Accurate Estimation of Running Temporal Parameters Using Foot-Worn Inertial Sensors. *Front. Physiol.* **2018**, *9*, 610. [CrossRef]
19. Al Haddad, H.; Méndez-Villanueva, A.; Torreño, N.; Munguía-Izquierdo, D.; Suárez-Arrones, L. Variability of GPS-derived running performance during official matches in elite professional soccer players. *J. Sports Med. Phys. Fit.* **2017**, *58*, 1439–1445. [CrossRef]
20. Hands, D.E.; Janse de Jonge, X. Current time-motion analyses of professional football matches in top-level domestic leagues: A systematic review. *Int. J. Perform. Anal. Sport* **2020**, *20*, 747–765. [CrossRef]
21. Di Salvo, V.; Baron, R.; Tschan, H.; Calderon Montero, F.J.; Bachl, N.; Pigozzi, F. Performance Characteristics According to Playing Position in Elite Soccer. *Int. J. Sports Med.* **2007**, *28*, 222–227. [CrossRef] [PubMed]
22. Sarmento, H.; Clemente, F.M.; Harper, L.D.; Costa, I.T.D.; Owen, A.; Figueiredo, A.J. Small sided games in soccer—a systematic review. *Int. J. Perform. Anal. Sport* **2018**, *18*, 693–749. [CrossRef]
23. Di Salvo, V.; Baron, R.; González-Haro, C.; Gormasz, C.; Pigozzi, F.; Bachl, N. Sprinting analysis of elite soccer players during European Champions League and UEFA Cup matches. *J. Sports Sci.* **2010**, *28*, 1489–1494. [CrossRef] [PubMed]
24. Gregson, W.; Drust, B.; Atkinson, G.; Salvo, V.D. Match-to-Match Variability of High-Speed Activities in Premier League Soccer. *Int. J. Sports Med.* **2010**, *31*, 237–242. [CrossRef]
25. Mohr, M.; Krustrup, P.; Bangsbo, J. Match performance of high-standard soccer players with special reference to development of fatigue. *J. Sports Sci.* **2003**, *21*, 519–528. [CrossRef]
26. Miguel, M.; Oliveira, R.; Brito, J.P.; Loureiro, N.; García-Rubio, J.; Ibáñez, S.J. External Match Load in Amateur Soccer: The Influence of Match Location and Championship Phase. *Healthcare* **2022**, *10*, 594. [CrossRef]
27. Saward, C.; Morris, J.G.; Nevill, M.E.; Nevill, A.M.; Sunderland, C. Longitudinal development of match-running performance in elite male youth soccer players. *Scand. J. Med. Sci. Sports* **2015**, *26*, 933–942. [CrossRef]
28. Varley, M.C.; Gregson, W.; McMillan, K.; Bonanno, D.; Stafford, K.; Modonutti, M.; Di Salvo, V. Physical and technical performance of elite youth soccer players during international tournaments: Influence of playing position and team success and opponent quality. *Sci. Med. Footb.* **2016**, *1*, 18–29. [CrossRef]
29. Buchheit, M.; Mendez-Villanueva, A.; Simpson, B.M.; Bourdon, P.C. Repeated-Sprint Sequences During Youth Soccer Matches. *Int. J. Sports Med.* **2010**, *31*, 709–716. [CrossRef]
30. Pillitteri, G.; Thomas, E.; Battaglia, G.; Navarra, G.A.; Scardina, A.; Gammino, V.; Ricchiari, D.; Bellafiore, M. Validity and Reliability of an Inertial Sensor Device for Specific Running Patterns in Soccer. *Sensors* **2021**, *21*, 7255. [CrossRef]
31. Riboli, A.; Coratella, G.; Rampichini, S.; Cé, E.; Esposito, F. Area per player in small-sided games to replicate the external load and estimated physiological match demands in elite soccer players. *PLoS ONE* **2020**, *15*, e0229194. [CrossRef] [PubMed]
32. Riboli, A.; Olthof, S.B.; Esposito, F.; Coratella, G. Training elite youth soccer players: Area per player in small-sided games to replicate the match demands. *Biol. Sport* **2022**, *39*, 579–598. [CrossRef] [PubMed]

33. Di Prampero, P.E.; Fusi, S.; Sepulcri, L.; Morin, J.B.; Belli, A.; Antonutto, G. Sprint running: A new energetic approach. *J. Exp. Biol.* **2005**, *208*, 2809–2816. [CrossRef]
34. Minetti, A.E.; Moia, C.; Roi, G.S.; Susta, D.; Ferretti, G. Energy cost of walking and running at extreme uphill and downhill slopes. *J. Appl. Physiol.* **2002**, *93*, 1039–1046. [CrossRef] [PubMed]
35. Coutts, A.J.; Duffield, R. Validity and reliability of GPS devices for measuring movement demands of team sports. *J. Sci. Med. Sport* **2010**, *13*, 133–135. [CrossRef]
36. Impellizzeri, F.M.; Marcora, S.M.; Coutts, A.J. Internal and External Training Load: 15 Years On. *Int. J. Sports Physiol. Perform.* **2019**, *14*, 270–273. [CrossRef]
37. Ravé, G.; Granacher, U.; Boullosa, D.; Hackney, A.C.; Zouhal, H. How to Use Global Positioning Systems (GPS) Data to Monitor Training Load in the "Real World" of Elite Soccer. *Front. Physiol.* **2020**, *11*, 944. [CrossRef]
38. Abbott, W.; Brickley, G.; Smeeton, N.J. Physical demands of playing position within {English} {Premier} {League} academy soccer. *J. Hum. Sport Exerc.* **2018**, *13*, 285–295. [CrossRef]
39. Ademović, A. Differences In the Quantity and Intesity of Playing In Elite Soccer Players of Different Position in the Game. *Homosporticus* **2016**, *18*, 26–31.
40. Andrzejewski, M.; Konefał, M.; Chmura, P.; Kowalczuk, E.; Chmura, J. Match outcome and distances covered at various speeds in match play by elite German soccer players. *Int. J. Perform. Anal. Sport* **2016**, *16*, 817–828. [CrossRef]
41. Aslan, A.; Acikada, C.; Güvenç, A.; Gören, H.; Hazir, T.; Ozkara, A. Metabolic demands of match performance in young soccer players. *J. Sports Sci. Med.* **2012**, *11*, 170–179.
42. Mendez-Villanueva, A.; Buchheit, M.; Simpson, B.; Bourdon, P.C. Match Play Intensity Distribution in Youth Soccer. *Int. J. Sports Med.* **2013**, *34*, 101–110. [CrossRef]
43. Di Salvo, V.; Gregson, W.; Atkinson, G.; Tordoff, P.; Drust, B. Analysis of High Intensity Activity in Premier League Soccer. *Int. J. Sports Med.* **2009**, *30*, 205–212. [CrossRef]
44. Ingebrigtsen, J.; Dalen, T.; Hjelde, G.H.; Drust, B.; Wisløff, U. Acceleration and sprint profiles of a professional elite football team in match play. *Eur. J. Sport Sci.* **2014**, *15*, 101–110. [CrossRef]
45. Ade, J.; Fitzpatrick, J.; Bradley, P.S. High-intensity efforts in elite soccer matches and associated movement patterns, technical skills and tactical actions. Information for position-specific training drills. *J. Sports Sci.* **2016**, *34*, 2205–2214. [CrossRef]
46. Orendurff, M.S.; Walker, J.D.; Jovanovic, M.; Tulchin, K.L.; Levy, M.; Hoffmann, D.K. Intensity and Duration of Intermittent Exercise and Recovery During a Soccer Match. *J. Strength Cond. Res.* **2010**, *24*, 2683–2692. [CrossRef]
47. Bangsbo, J. Energy demands in competitive soccer. *J. Sports Sci.* **1994**, *12*, S5–S12. [CrossRef]
48. Gaudino, P.; Iaia, F.M.; Alberti, G.; Strudwick, A.J.; Atkinson, G.; Gregson, W. Monitoring Training in Elite Soccer Players: Systematic Bias between Running Speed and Metabolic Power Data. *Int. J. Sports Med.* **2013**, *34*, 963–968. [CrossRef]
49. Bangsbo, J.; Nørregaard, L.; Thorsø, F. Activity profile of competition soccer. *Can. J. Sport Sci. = J. Can. des Sci. du Sport* **1991**, *16*, 110–116.
50. Aquino, R.; Carling, C.; Vieira, L.H.P.; Martins, G.; Jabor, G.; Machado, J.; Santiago, P.; Garganta, J.; Puggina, E. Influence of Situational Variables, Team Formation, and Playing Position on Match Running Performance and Social Network Analysis in Brazilian Professional Soccer Players. *J. Strength Cond. Res.* **2020**, *34*, 808–817. [CrossRef]
51. Trewin, J.; Meylan, C.; Varley, M.; Cronin, J. The influence of situational and environmental factors on match-running in soccer: A systematic review. *Sci. Med. Footb.* **2017**, *1*, 183–194. [CrossRef]
52. Altmann, S.; Neumann, R.; Woll, A.; Härtel, S. Endurance Capacities in Professional Soccer Players: Are Performance Profiles Position Specific? *Front. Sports Act. Living* **2020**, *2*, 549897. [CrossRef] [PubMed]
53. Rampinini, E.; Coutts, A.J.; Castagna, C.; Sassi, R.; Impellizzeri, F.M. Variation in top level soccer match performance. *Int. J. Sports Med.* **2007**, *28*, 1018–1024. [CrossRef]
54. Dalen, T.; Jørgen, I.; Gertjan, E.; Havard, H.G.; Ulrik, W. Player Load, Acceleration, and Deceleration During Forty-Five Competitive Matches of Elite Soccer. *J. Strength Cond. Res.* **2016**, *30*, 351–359. [CrossRef] [PubMed]
55. Buchheit, M.; Samozino, P.; Glynn, J.A.; Michael, B.S.; Al Haddad, H.; Mendez-Villanueva, A.; Morin, J.-B. Mechanical determinants of acceleration and maximal sprinting speed in highly trained young soccer players. *J. Sports Sci.* **2014**, *32*, 1906–1913. [CrossRef]
56. Little, T.; Williams, A.G. Specificity of acceleration, maximum speed, and agility in professional soccer players. *J. Strength Cond. Res.* **2005**, *19*, 76–78.
57. Varley, M.C.; Aughey, R.J. Acceleration profiles in elite Australian soccer. *Int. J. Sport. Med.* **2013**, *34*, 34–39. [CrossRef]
58. Vanrenterghem, J.; Nedergaard, N.J.; Robinson, M.A.; Drust, B. Training Load Monitoring in Team Sports: A Novel Framework Separating Physiological and Biomechanical Load-Adaptation Pathways. *Sports Med.* **2017**, *47*, 2135–2142. [CrossRef]
59. Hader, K.; Mendez-Villanueva, A.; Palazzi, D.; Ahmaidi, S.; Buchheit, M. Metabolic Power Requirement of Change of Direction Speed in Young Soccer Players: Not All Is What It Seems. *PLoS ONE* **2016**, *11*, e0149839. [CrossRef]
60. Riboli, A.; Esposito, F.; Coratella, G. Small-Sided Games in Elite Football: Practical Solutions to Replicate the 4-min Match-Derived Maximal Intensities. *J. Strength Cond. Res.* **2022**, *37*, 366–374. [CrossRef]
61. Bourgeois, F.A.; McGuigan, M.R.; Gill, N.D.; Gamble, P. Physical characteristics and performance in change of direction tasks: A brief review and training considerations. *J. Aust. Strength Cond.* **2017**, *25*, 104–117.

62. DeWeese, B.H.; Nimphius, S. Program design and technique for speed and agility training. In *Essentials of Strength Training and Conditioning*; Human Kinetics Publishers: Champaign, IL, USA, 2018.
63. Manzi, V.; Annino, G.; Savoia, C.; Caminiti, G.; Padua, E.; Masucci, M.; D'Onofrio, R.; Iellamo, F. Relationship between aerobic fitness and metabolic power metrics in elite male soccer players. *Biol. Sport* **2022**, *39*, 599–606. [CrossRef]

Journal of
*Functional Morphology
and Kinesiology*

MDPI

Review

The Metaverse: A New Challenge for the Healthcare System: A Scoping Review

Luca Petrigna [1,*] **and Giuseppe Musumeci** [1,2]

[1] Department of Biomedical and Biotechnological Sciences, Section of Anatomy, Histology and Movement Science, School of Medicine, University of Catania, Via S. Sofia n°97, 95123 Catania, Italy
[2] Research Center on Motor Activities (CRAM), University of Catania, Via S. Sofia n°97, 95123 Catania, Italy
* Correspondence: luca.petrigna@unict.it

Abstract: (1) Background: The metaverse is now a reality, and it interests the scientific community, the educational setting, and medical care. Considering the number of people in front of screens, especially children and adolescents, the metaverse could and should become a place of health promotion. Consequently, the objective of the present study was to review the current literature to detect articles that connected the metaverse with prevention and treatment, education and training, and research setting. (2) Methods: Articles were searched on Pubmed, Web of Science, and Scopus, including English-written papers published until 12 August 2022. They were screened against the eligibility criteria and discussed narratively. (3) Results: The literature published is poor; only 21 articles were included, and 11 of them were added in a second moment. These articles were mainly reviews of the literature or editorials. The aspects related to this virtual world in terms of health prevention and the treatment of clinical conditions, education and training, and research have been narratively discussed. (4) Conclusions: The metaverse could be considered a useful instrument to arrive easily and quickly to the population. Given its importance, today, different studies and investments are required to develop proper health promotion programs that are feasible and valid in the metaverse.

Keywords: virtual reality; augmented reality; lifelogging; mirror world; health

Citation: Petrigna, L.; Musumeci, G. The Metaverse: A New Challenge for the Healthcare System: A Scoping Review. *J. Funct. Morphol. Kinesiol.* **2022**, *7*, 63. https://doi.org/10.3390/jfmk7030063

Academic Editor: Helmi Chaabene

Received: 6 July 2022
Accepted: 24 August 2022
Published: 30 August 2022

1. Introduction

The term "metaverse" was introduced for the first time by Neal Stephenson in 1992, and the science fiction novel Snow Crash is about an immersive and alternative virtual reality, and the internet-connected universe [1] becomes a reality. The metaverse is an internet-based 3-dimensional (3D) virtual world where people conduct daily activities using avatars representing their "real" or imaginary themselves [2]. In a few words, a virtual space became the real world for an alternative life where avatars or digital profiles participate in social activities and in virtual cultural events but also have an economic life [2].

About the structure and technologies behind the metaverse, it can be divided into four different categories: augmented reality, lifelogging, mirror world, and virtual reality [2]:

- Augmented reality adds, in real-time, a digital graphic environment to an existing, physical, and real world. It uses glasses, lenses, or smartphones. In the metaverse, the idea is to superimpose further information on the real environment. Examples are Pokemon Go and 3D medical animations.
- Lifelogging is also an augmentation of the inner word. Different from augmented reality, smart devices are adopted to record daily lives on the internet. Examples are Instagram, Facebook, Twitter, and health monitors.
- A mirror world is a simulation of the real world. The real appearance, information, and structure are transferred to a virtual space, allowing the performance of activities through the internet or mobile applications. Examples are Google Maps or Earth,

educational spaces, such as "digital laboratories" and "virtual educational spaces," but also Zoom, Webex, Google Meet, and Teams.

- Virtual reality is a virtual online sophisticated 3D reality with avatars and an instant communication tool that simulates the inner world. The avatar can be personalized, and the cultural, physical, and social characteristics are different from reality. The avatar can communicate with other entities and achieves goals. Examples are online multiplayer video games, virtual hospitals, and consultation rooms.

The metaverse, therefore, can be considered a location in which the real world is augmented, connected, and replicated with virtual reality and, consequently, as another world [2]. It is also important to consider that for the digital native generation, the metaverse is and will be a space where they spend part of their daily life [2]. These are aspects that have to be considered, especially after the coronavirus disease of the 2019 pandemic that accelerated and implemented the evolution of the metaverse [2,3]. The pandemic situation limited real physical life [4]; consequently, activities such as education, medical care, fashion, shopping, performances, exhibitions, and tourism moved from only offline to be also virtual [2]. So, the metaverse is not only a place to get away from the busy life and enjoy the leisurely life [2], but according to us, it is a place where people will live part of their life using services and facilities. This connection with the metaverse is facilitated by new technologies that allow us to be part of this online world 24 h a day and everywhere. These devices are smartphones and smartwatches, with the last allowing healthcare education, physical activity monitoring, but also the possibility of the self-management of chronic diseases and nursing or home-based care [5].

Diverse industries, private companies, and organizations, from social communications to fashion, from high-tech to business, and from art to real estate, are actively investing and creating virtual entities in the metaverse. The healthcare sector is now starting to deal with the metaverse, and the potentiality of this virtual world for the prevention and treatment of clinical conditions, education and training, and research is limitless. For these reasons and the objective of the present study was to understand the progress of the scientific literature on the metaverse in the healthcare sector and eventually to propose some feedback in terms of health prevention and the treatment of clinical conditions, education and training, and research. Furthermore, the practical utility of the metaverse has also been considered. It is, consequently, important and interesting to review the literature published on the metaverse related to public health to have a clear idea of what it has done and what it is required to do.

2. Materials and Methods

The manuscript partially followed the preferred reporting for systematic reviews and meta-analyzes for Scoping Reviews (PRISMA-ScR) checklist and explanation [5]. PRISMA checklist is included in the Supplementary Material. The manuscript was not previously recorded on databases such as PROSPERO.

2.1. Search Strategy

The studies were collected through a screening of the electronic databases PubMed (NLM), the Web of Sciences, and Scopus, and the manuscripts were collected if published before 12 August 2022. The only keyword adopted was metaverse to include as many articles as possible in the review.

2.2. Eligibility Criteria

No eligibility restrictions were adopted for the population investigated, the intervention adopted, the comparison parameters, the outcomes of the studies, and the study design. The only inclusion criteria were related to the topic that had to be on the metaverse and public health. Only original, peer-reviewed, and English-written manuscripts were included independently of the origin country. No letter to the editor, conference papers, proceedings, or abstracts, while other manuscript typologies were included.

2.3. Data Sources, Studies Sections, and Data Extraction

Two investigators performed the study collection and screening. The first step of the screening was the removal of duplicate screenings, which were performed by title, abstract, and full text. Furthermore, in a second step, the references of the included studies were screened to include the review of other studies on this topic.

The main results were summarized in a table where information related to the year of publication, journal, study typology, and the main objective were collected. The results were narratively analyzed and discussed.

3. Results

A total of 976 (Pubmed: 67; Web of Science: 380; Scopus: 529) studies were detected after the screening of the three databases. Most of the excluded articles were related to other topics. Only 10 articles were included after the eligibility criteria selection. From the references of the included studies, another 11 studies were included. The final number of the included studies was 21. The screening process is presented in Figure 1.

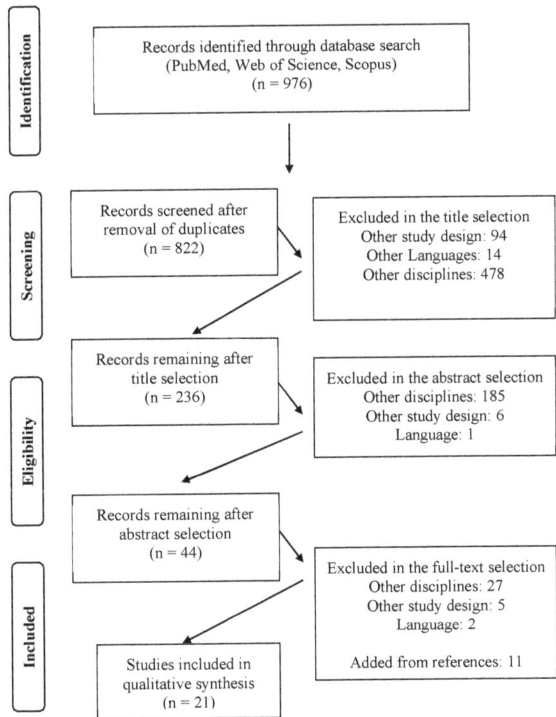

Figure 1. Flow chart of the selection criteria process of the included studies.

Characteristics of the Included Studies

Despite no limits being proposed for the year of publication, 1 study was published in 2018 [6] and 1 in 2019 [7]. Five studies were published in 2020, 4 studies in 2021, and 10 studies in 2022. A total of 14 studies were reviewed in the literature, and only one of these studies performed was also a meta-analysis. Five studies were Editorials, one study a prospective, and one study a Global Spotlight. The objectives of the included studies were different, but it was possible to categorize the results in (a) health prevention and the treatment of clinical conditions; (b) education and training, and (c) research. More details are presented in Table 2.

Table 1. Characteristics of the included studies.

Author	Year	Journal	Study Typology	Main Objective
Abu-Elezz et al. [8]	2020	International Journal of Medical Informatics	Review	Explore and categorise the benefits and threats of blockchain technology application in a healthcare system
All-Jaroodi et al. [9]	2020	IEEE Access	Review	Define Health 4.0 and discuss advanced potential Health 4.0 applications
Almarzouqi et al. [10]	2022	Digital Object Identifier	Review	Evaluate students' perception of the application of the metaverse for medical-educational purposes
Aziz et al. [6]	2018	Journal of Health & Medical Informatics	Review	Summarizes the current state of knowledge of virtual reality simulation in healthcare
Chapman et al. [11]	2022	SAGE	Editorial	The metaverse is associated with training in lung cancer surgery
Chen et al. [12]	2020	Journal of medical Internet research	Review, meta-analysis	Evaluate the effectiveness of virtual reality in nursing education in the areas of knowledge, skills, satisfaction, confidence, and performance time
Javaid et al. [13]	2020	Clinical Epidemiology and Global Health	Review	Find how the metaverse is going to solve a medical-related problem in saving the life of the patient and what are the significant applications
Kye et al. [2]	2021	Journal of educational evaluation for health professions	Review	Define the 4 types of the metaverse and explain the potential and limitations of its educational applications
Koo et al. [14]	2021	J Educ Eval Health Prof	Editorial	Training in lung cancer surgery through the metaverse
Krittanawong et al. [15]	2021	Canadian Journal of Cardiology	Review	Discuss recent advances and potential future directions for the application of blockchain and its integration with artificial intelligence in cardiovascular medicine
Krittanawong et al. [16]	2020	Nature reviews. Cardiology	Review	Discuss integration of blockchain with artificial intelligencedata-centric analysis and information flow, its limitations, and potential cardiovascular applications
Liu et al. [4]	2022	Int. J. Environ. Res. Public Health	Review	Explore VR in aiding therapy, providing a potential guideline for futures application in healthcare towards Health 4.0
Mesko et al. [17]	2022	European Heart Journal	Global Spotlights	It was associated with the metaverse with cardiovascular health
Ramesh et al. [18]	2022	Indian J Ophthalmol	Editorial	Presentation of the 4D ophthalmic anatomical and pathological metaverse
Schuelke et al. [7]	2019	Nursing administration quarterly	Review	Report on an innovative care system and the effects of this model have on patient satisfaction, patient quality metrics, and financial metrics

Table 2. Characteristics of the included studies.

Author	Year	Journal	Study Typology	Main Objective
Skalidis et al. [19]	2022	Trends in Cardiovascular Medicine	Review	Analysis of the applications of the metaverse and how it can be implemented in cardiovascular medicine
Tan et al. [20]	2022	Asia-Pacific Academy of Ophthalmology	Perspective	Analysis opportunities and challenges of the metaverse in ophthalmology
Usmani et al. [21]	2022	General Psychiatry	Review	Explore the applications of the metaverse on mental health
Wiederhold et al. [3]	2022	Cyberpsychology, behavior, and social networking	Editorial	Application of the metaverse in the health care setting
Yeung et al. [22]	2021	Journal of medical internet research	Review	Association of virtual reality and augmented reality with medicine
Zeng et al. [23]	2022	Asia-Pacific Journal of Oncology Nursing	Editorial	Application of the metaverse in cancer care

4. Discussion

The findings suggest that the literature on this topic is limited to a few reviews of the literature and editorials. The included studies are recent, and the metaverse was adopted with different scopes requiring further investigations in the near future. Despite these limitations, the metaverse can be applied in the health prevention and treatment of clinical conditions; it is feasible in the education and training setting, and researchers can use this tool to make the studies faster and with bigger and worldwide samples. These aspects are deeply presented in the paragraphs below.

4.1. Metaverse for Prevention and Treatment

The first aspect that we wanted to analyze is the relation between the metaverse and the prevention and treatment of disease. A bibliometric analysis of virtual reality and augmented reality found that the metaverse can be adopted for diagnostic and surgical procedures and rehabilitation on pain, stroke, anxiety, depression, fear, cancer, and neurodegenerative disorders with satisfying results [23]. Specific, for cancer care, artificial intelligence technologies can be a tool to prevent cancer and diagnose, treat, and rehabilitate patients [21,24].

A widely adopted concept is Health 4.0, and it integrates innovative technologies with health care [4,10]. Examples of Health 4.0 are the Internet of Health Things, medical cyberphysical systems, health cloud or fog, big data analytics, machine learning, blockchain, and smart algorithms [10] but also virtual reality [4]. This allows us to monitor the population but also to educate involving people in community activities. Digital innovations can be adopted as an alternative model of care delivery, and indeed, the possibility to create avatars allows consultations and personalized care [21]. In the metaverse, physicians could visit their patients in a 3D virtual clinic using telemedicine services and home-based devices such as wearable sensors and smartphone applications to monitor their health status [18].

Different devices can be adopted to monitor health conditions directly at home, connecting real life with the virtual world. The monitoring of the clinical aspect and, consequently, the health of the person who is distant can be adopted by 12-lead electrocardiograms for the heart, blood pressure instruments to evaluate the cardiovascular systems, oxygen saturation meters for the cardio-respiratory system, and blood glucose calculators, ideally for people with diabetes [20]. Related to the evaluation of physical performance, also through the web, widely adopted are heart frequency monitors [20]. Another tool widely adopted, especially in the last years, is the smartwatch that integrates heart frequency, blood saturation, pedometers, and accelerometers, but also Global Positioning System (GPS) data allow the monitoring of health status and physical performance, but also the management of chronic disease [5,25,26]. These smartwatches are often connected to the smartphone, and the smartphone is connected to communities where people can compare, in real-time, their data with other users and accept challenges. The potentiality of smartwatches in health promotion programs is huge. Indeed, through real-time monitoring, the possibility of being part of an online community and being guided by experts around the world increases the possibility of attending fitness programs and adopting a healthy lifestyle [27]. These smartwatches could be an instrument to make the person alive in the metaverse. Furthermore, the 24 h a day monitoring of health parameters allows prompt prevention or intervention in case of problems such as atrial fibrillation [28], also helping the service provider to improve the security of the intervention. Along with monitoring, another important aspect to consider is that, in the metaverse, a virtual and artificial intelligence-based avatar or an agent could provide personalized feedback and motivations, and, in this way, the intervention can become more effective in promoting behavior change [29]. These considerations can also be done considering that the avatars, thanks to the new technology, react realistically in their speech, facial expressions, and body language [30]. Virtual reality could be adopted to create pre-operative planning by analyzing the situation in 3D and asking for the opinions of other experts connected through the internet [12,18]. Furthermore, virtual reality can be integrated with different therapies, and it is an effective intervention for various medical conditions with advantages in terms of customization, compliance, cost, accessibility, motivation, and convenience [4]. Virtual reality can also be adopted as a rehabilitation technique, and an example is mirror therapy, which is a way to create a visual illusion that an impaired limb (an example is after a stroke) is correctly working to re-educate the movement of the cognitive patterns [31].

One last aspect to consider is the possibility of creating an avatar that can act as a "virtual nurse" to direct and monitor care and interact with the patient educating him or the staff around him, but also supervise and monitor, in real-time, the quality/patient safety surveillance, physician activity, and admission and discharge activities [8]. On one side, the metaverse can serve as a transitional stage before real-world experiences [3] with healthcare providers that can accompany patients into specific individualized environments enhancing the efficacy of treatment [3]. On the other side, for chronic diseases such as hypertension, diabetes, obesity, and some mental health conditions, virtual care models with psychological group support programs could be a valid intervention, while remote virtual nursing care with robotic end-user delivery units could also be of help [12]. Regarding mental healthcare, the metaverse with an anonymous virtual realm could be a comfortable place to share stories and other issues with professionals [22]. A representation of this alternative doctor–patient relationship is in Figure 2.

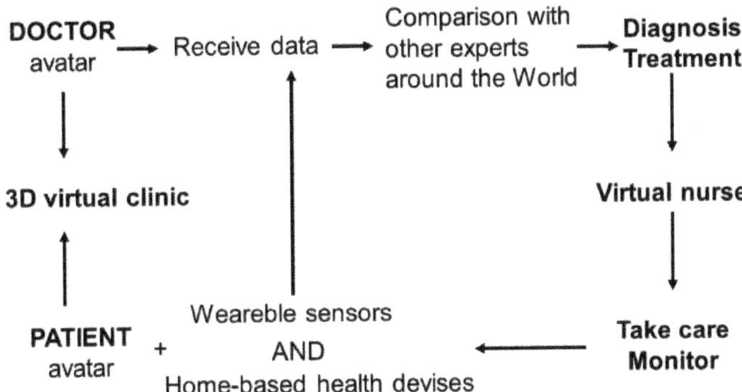

Figure 2. Representation of the possible virtual relationship between doctors and patients.

Health promotion can be performed in teaching rooms, and how it is the education of students, the same method can also be adopted for this aspect. However, health promotion programs also mean allowing people to have a space to train. We can imagine virtual rooms in which people can meet with other people and train together. Something similar, just existing and adopted, is the Peloton. It is a virtual environment for bicycling [32]. Peloton is a sports simulator that uses virtual reality, and now, an interactive video where people can compete or exercise together in shared virtual spaces. Users can walk or run on treadmills or by pedaling on stationary bicycles [33]. Obviously, people need exercise equipment that is connected to a local computer through sensors and force feedback controls [33]. In this way, people can train in nature and with other people, inside their houses, offices, or inside the International Space Station where training is a fundamental element [34]. People in their avatar form will be dressed properly, but at home, alone, and they can wear a proper gym suit but also pajamas. The sensors on the equipment will monitor the training of the user and the avatar instructor will guide the users to reach their goals [33]. Feedback could also be the modification of the inclination of the treadmills or the pedaling resistance of their bicycles [33]. People from their homes, offices, or exercise clubs can participate in exercise sessions or compete [33]. According to us, this is an easy way to bring people to move daily, also from their houses, decreasing sedentary behaviors, fundamental aspect nowadays.

4.2. Metaverse Education and Training

As the metaverse began to be introduced into present life rapidly, some metaverse applications have already been adopted in education [2], but further improvements are necessary. In an educational setting, virtual reality is the most adopted technology of the metaverse [2,11]. The advantage of virtual reality is that it can be accessed from anywhere, regardless of distance or space [2]. This educational methodology can be adopted by a professor of an important university that can teach people the same methodology in its virtual reality and can also be used for training, for example, its application in surgery, cardiology, and neurology [14]. The overlap of abstract visuals and virtual objects in the context of the real world can be useful for visualizing virtual internal organs and structures of a real body. In this way, it is possible to visualize the internal part of a body directly on the t-shirt of an alive person, or professors could teach aspiring cardiologists the inner workings of the heart and the cardiovascular system in 3D [18]. Some authors used an extended reality technology to create a holographic museum of anatomical structures, such as the eyeball, cerebral venous system, cerebral arterial system, cranial nerves, and various parts of the brain, making it a useful tool in ophthalmic teaching [19]. Like virtual reality, augmented reality can also be adopted to teach during clinical education [11,15]. Both these technologies and virtual rooms allow people to reach each other all around the world and also in remote locations, allowing the standardization of the education of people. This could decrease the discrepancy in the education of future medical students. Virtual reality can be adopted to provide a broad range of training to medical care professionals in anatomy instructions and surgery simulations that can simulate and control different situations that are difficult to duplicate in real life [7].

Another feasible and cheap way to educate people with health education programs organized by experts in the field, universities, or governments is through online platforms. This is a distance learning methodology that can reach people all around the world, 24 h a day, and can be registered while remaining online for an indefinite period of time. Distance learning can also include practical elements such as physical and behavioral education programs.

4.3. Metaverse and Research Application

An online system such as the metaverse could provide the possibility of collecting a huge amount of personal health information, allowing the creation of big data and machine learning systems that could help the health care system and research [35]. The data recorded could create national or international monitoring and surveillance systems that researchers could use for their studies, both to obtain data and to compare their results. These digital data could be collected directly from the consumer or from the services or wearables, and they can be shared with a physician, medical professionals, or researchers [18].

The metaverse is an online world in evolution, and the role of behavioral healthcare providers and researchers in guiding its formation and conducting unbiased research is important [30]. For this reason, it is fundamental to create a solid base of where to start and to try to understand this topic as soon as possible and as well as possible to take the right way. A graphic representation of how the data from people can arrive at researchers and public organizations is in Figure 3.

Another important aspect to consider is the possibility of using extended reality to simulate, within a laboratory, a different location, such as nature or a city, but also a particular setting or a multiple task exercise [36] to stimulate the cognitive system. This kind of technology is just adopted, and examples are glasses or screens. Some manufacturing companies are just selling laboratory instruments, such as the treadmill, associated with screens or special glasses to create an extended reality.

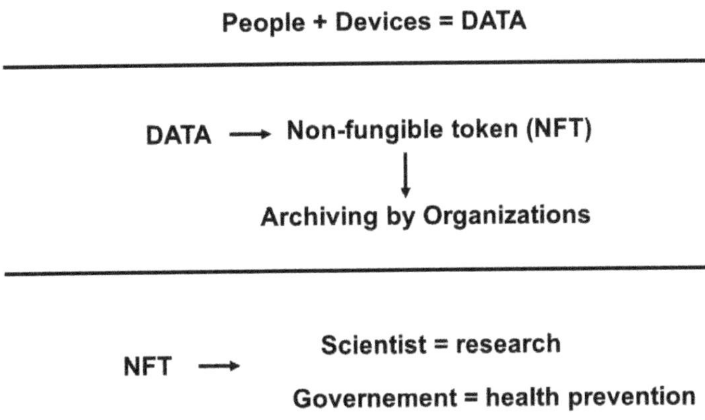

People + Devices = DATA

DATA ⟶ **Non-fungible token (NFT)**

Archiving by Organizations

NFT ⟶ **Scientist = research**

Governement = health prevention

Figure 3. Graphic representation of the passage of the data from people to the researchers and public organizations.

4.4. Limit of the Metaverse

The main limits of the metaverse are related to data management and privacy, cyber-security risks, potential barriers related to access (a lack of internet connectivity), and users with low vision [21]. The metaverse could be a dangerous place with possible new vicious and sophisticated crimes, especially related to stealing personal data [2].

The first aspect to consider is the issues related to the management and storing of data. The commercial interest is huge, and people could have limited power in controlling their data sharing with whom and under what conditions [35]. Consequently, it becomes fundamental to solve the problems related to privacy and security, and technical, legislative, and regulatory problems [20,30]. Also, healthcare leaders have expressed concerns about privacy, ethics, and safety as healthcare moves online [3], making it necessary to enhance data security and centralized regulatory oversight [16]. A method to organize and control access to a complex network of data sets is through blockchain [3,16]. Blockchain is a viable technology that can improve healthcare data sharing and the storing system, owing to its decentralization, immutability, transparency, traceability features, and privacy [3,9]. Blockchain networks integrated with artificial intelligence could help in the personalization of cardiovascular medicine by (a) increasing the data available for developing and training artificial intelligence, (b) sharing algorithms for generalization, and (c) decentralizing databases and incentivizing solutions that improve outcomes [16,17]. Consequently, blockchain could be a new way of encrypting patient data and enforcing compliance with medical standards in practices and processes [20]. The application of blockchain to cardiovascular medicine is still in the beginning, and some concerns about the implementation exist [16,17]. Healthcare organizations are hesitant to adopt blockchain technology due to threats such as security, authorization, and interoperability issues and the lack of technical skills related to blockchain technology [9]. Also, non-fungible tokens could become helpful. Indeed, they allow digital ownership, and they can help incentivize a more democratized, transparent, and efficient system for patients to control their data sharing [35]. In this way, the health system could have the whole medical history, disease, medication, and allergies stored in the patients' own personalized non-fungible token that only they and their doctor can access (or any other individual that the patient decides to give access to) [20].

The second aspect to consider is the physical relationship. It currently remains much easier to communicate with patients in person, through a phone, or a social media channel than to envision something that seems like a far-fetched science fiction idea [18]. A second limitation is that the metaverse is expected to be actively used in medical and nursing education or education for residents and students [15]. However, even if virtual reality can effectively improve knowledge in nursing education, it is not more effective than other education methods in the areas of skills, satisfaction, confidence, and performance time [13]. Social connections in the metaverse are weaker than interactions in the real world [2]. Furthermore, it is still an expensive technology.

The last aspect is the possible loss of freedom in the future. This is how reality should be, and it is the advantage of the actual metaverse. Consequently, the main issue is if the metaverse must be administrated by the public or under the control of multi-millionaire companies that are developing this alternative reality. Probably, the metaverse should be a representation of the reality with the public and private that will coexist. This means that the government should have to start to invest in this alternative world to guarantee public spaces, educational programs, and public events. One last aspect to consider, both by the public and private administration, is to guarantee personal data privacy with a proper constitution.

The limits of the study are mainly related to the poor quality of the included studies. This suggests that the topic is new and innovative, requiring future investigations. A second limitation of the study is related to the impossibility of having and analyzing real data but only theoretical concepts. Future studies should deeply investigate the feasibility of a health-metaverse in which people can learn and be guided in healthy behaviors. These studies should transform the theoretical concept in original research.

5. Conclusions

The metaverse cannot substitute the real world; physical and eye contact, facial expressions, and gestures are essential elements in the healthcare world. However, the metaverse can be considered a tool to improve the quality of the health care system in terms of intervention and treatment, the education of people all over the world, guaranteeing standardized training, and helping the research to create world databases. Finally, considering the time spent by the young population in front of a screen, the metaverse could be a place where they can also start to practice sport and learn something. Sports centers and associations could enrich those citizens who wish to train differently, at home, and when they prefer, increasing the possibilities to be physically active.

Supplementary Materials: The following are available online at https://www.mdpi.com/article/10.3390/jfmk7030063/s1, Figure S1: referred Reporting Items for Systematic reviews and Meta-Analyses extension for Scoping Reviews (PRISMA-ScR) Checklist.

Author Contributions: L.P. and G.M. equally worked on this manuscript. All authors have read and agreed to the published version of the manuscript.

Funding: This work was funded by the University Research Project Grant (PIACERI Found—NATURE-OA—2020–2022), the Department of Biomedical and Biotechnological Sciences (BIOME-TEC), and the University of Catania, Italy.

Institutional Review Board Statement: Not applicable.

Informed Consent Statement: Not applicable.

Data Availability Statement: Not applicable.

Conflicts of Interest: The authors declare no conflict of interest.

References

1. Stephenson, N. *Snow Crash: A Novel*; Spectra: New York, NY, USA, 1992.
2. Kye, B.; Han, N.; Kim, E.; Park, Y.; Jo, S. Educational applications of metaverse: Possibilities and limitations. *J. Educ. Eval. Health Prof.* **2021**, *18*, 30–32. [CrossRef] [PubMed]
3. Wiederhold, B.K. Metaverse Games: Game Changer for Healthcare? *Cyberpsychol. Behav. Soc. Netw.* **2022**, *25*, 267–269. [CrossRef] [PubMed]
4. Liu, Z.; Ren, L.; Xiao, C.; Zhang, K.; Demian, P. Virtual reality aided therapy towards health 4.0: A two-decade bibliometric analysis. *Int. J. Environ. Res. Public Health* **2022**, *19*, 1525. [CrossRef]

5. King, C.E.; Sarrafzadeh, M. A survey of smartwatches in remote health monitoring. *J. Healthc. Inform. Res.* **2018**, *2*, 1–24. [CrossRef] [PubMed]
6. Tricco, A.C.; Lillie, E.; Zarin, W.; O'Brien, K.K.; Colquhoun, H.; Levac, D.; Moher, D.; Peters, M.D.J.; Horsley, T.; Weeks, L.; et al. PRISMA Extension for Scoping Reviews (PRISMA-ScR): Checklist and Explanation. *Ann. Intern. Med.* **2018**, *169*, 467–473. [CrossRef]
7. Aziz, H.A. Virtual reality programs applications in healthcare. *J. Health Med. Inform.* **2018**, *9*, 305. [CrossRef]
8. Schuelke, S.; Aurit, S.; Connot, N.; Denney, S. Virtual nursing: The new reality in quality care. *Nurs. Adm. Q.* **2019**, *43*, 322–328. [CrossRef]
9. Abu-Elezz, I.; Hassan, A.; Nazeemudeen, A.; Househ, M.; Abd-Alrazaq, A. The benefits and threats of blockchain technology in healthcare: A scoping review. *Int. J. Med. Inform.* **2020**, *142*, 104246. [CrossRef]
10. Al-Jaroodi, J.; Mohamed, N.; Abukhousa, E. Health 4.0: On the way to realizing the healthcare of the future. *IEEE Access* **2020**, *8*, 211189–211210. [CrossRef]
11. Almarzouqi, A.; Aburayya, A.; Salloum, S.A. Prediction of User's Intention to Use Metaverse System in Medical Education: A Hybrid SEM-ML Learning Approach. *IEEE Access* **2022**, *10*, 43421–43434. [CrossRef]
12. Chapman, J.R.; Wang, J.C.; Wiechert, K. Into the Spine Metaverse: Reflections on a future Metaspine (Uni-) verse. *Glob. Spine J.* **2022**, *12*, 545–547. [CrossRef]
13. Chen, F.-Q.; Leng, Y.-F.; Ge, J.-F.; Wang, D.-W.; Li, C.; Chen, B.; Sun, Z.-L. Effectiveness of virtual reality in nursing education: Meta-analysis. *J. Med. Internet Res.* **2020**, *22*, e18290. [CrossRef] [PubMed]
14. Javaid, M.; Haleem, A. Virtual reality applications toward medical field. *Clin. Epidemiol. Glob. Health* **2020**, *8*, 600–605. [CrossRef]
15. Koo, H. Training in lung cancer surgery through the metaverse, including extended reality, in the smart operating room of Seoul National University Bundang Hospital, Korea. *J. Educ. Eval. Health Prof.* **2021**, *18*, 33. [CrossRef] [PubMed]
16. Krittanawong, C.; Aydar, M.; Virk, H.U.H.; Kumar, A.; Kaplin, S.; Guimaraes, L.; Wang, Z.; Halperin, J.L. Artificial Intelligence–Powered Blockchains for Cardiovascular Medicine. *Can. J. Cardiol.* **2021**, *38*, 185–195. [CrossRef] [PubMed]
17. Krittanawong, C.; Rogers, A.; Aydar, M.; Choi, E.; Johnson, K.; Wang, Z.; Narayan, S.M. Integrating blockchain technology with artificial intelligence for cardiovascular medicine. *Nat. Reviews. Cardiol.* **2020**, *17*, 1–3. [CrossRef]
18. Mesko, B. The promise of the metaverse in cardiovascular health. *Eur. Heart J.* **2022**, *43*, 2647–2649. [CrossRef]
19. Ramesh, P.V.; Joshua, T.; Ray, P.; Devadas, A.K.; Raj, P.M.; Ramesh, S.V.; Ramesh, M.K.; Rajasekaran, R. Holographic elysium of a 4D ophthalmic anatomical and pathological metaverse with extended reality/mixed reality. *Indian J. Ophthalmol.* **2022**, *70*, 3116–3121. [CrossRef]
20. Skalidis, I.; Muller, O.; Fournier, S. CardioVerse: The Cardiovascular Medicine in the Era of Metaverse. *Trends Cardiovasc. Med.* **2022**. Available online: https://www.sciencedirect.com/science/article/pii/S1050173822000718 (accessed on 8 July 2022).
21. Tan, T.F.; Li, Y.; Lim, J.S.; Gunasekeran, D.V.; Teo, Z.L.; Ng, W.Y.; Ting, D.S. Metaverse and Virtual Health Care in Ophthalmology: Opportunities and Challenges. *Asia-Pac. J. Ophthalmol. (Phila. Pa.)* **2022**, *11*, 237–246. [CrossRef]
22. Usmani, S.S.; Sharath, M.; Mehendale, M. Future of mental health in the metaverse. *Gen. Psychiatry* **2022**, *35*, e100825. [CrossRef]
23. Yeung, A.W.K.; Tosevska, A.; Klager, E.; Eibensteiner, F.; Laxar, D.; Stoyanov, J.; Glisic, M.; Zeiner, S.; Kulnik, S.T.; Crutzen, R. Virtual and augmented reality applications in medicine: Analysis of the scientific literature. *J. Med. Internet Res.* **2021**, *23*, e25499. [CrossRef]
24. Zeng, Y.; Zeng, L.; Zhang, C.; Cheng, A.S. *The Metaverse in Cancer Care: Applications and Challenges*; Elsevier: Amsterdam, The Netherlands, 2022; p. 100111.
25. Jat, A.S.; Grønli, T.-M. Smart Watch for Smart Health Monitoring: A Literature Review. In Proceedings of the International Work-Conference on Bioinformatics and Biomedical Engineering, Maspalomas, Spain, 27–30 June 2022; pp. 256–268.
26. Debon, R.; Coleone, J.D.; Bellei, E.A.; De Marchi, A.C.B. Mobile health applications for chronic diseases: A systematic review of features for lifestyle improvement. *Diabetes Metab. Syndr. Clin. Res. Rev.* **2019**, *13*, 2507–2512. [CrossRef] [PubMed]
27. Amagai, S.; Pila, S.; Kaat, A.J.; Nowinski, C.J.; Gershon, R.C. Challenges in participant engagement and retention using mobile health apps: Literature review. *J. Med. Internet Res.* **2022**, *24*, e35120. [CrossRef] [PubMed]
28. Isakadze, N.; Martin, S.S. How useful is the smartwatch ECG? *Trends Cardiovasc. Med.* **2020**, *30*, 442–448. [CrossRef] [PubMed]
29. Taylor, L.; Ranaldi, H.; Amirova, A.; Zhang, L.; Ahmed, A.A.; Dibb, B. Using virtual representations in mHealth application interventions for health-related behaviour change: A systematic review. *Cogent Psychol.* **2022**, *9*, 2069906. [CrossRef]
30. Wiederhold, B.K. Ready (or Not) player one: Initial musings on the metaverse. *Cyberpsychol. Behav. Soc. Netw.* **2022**, *25*, 1–2. [CrossRef]
31. Zhang, Y.; Xing, Y.; Li, C.; Hua, Y.; Hu, J.; Wang, Y.; Ya, R.; Meng, Q.; Bai, Y. Mirror therapy for unilateral neglect after stroke: A systematic review. *Eur. J. Neurol.* **2022**, *29*, 358–371. [CrossRef]
32. Ensor, J.R.; Carraro, G.U. Peloton: A distributed simulation for the world wide web. *Simul. Ser.* **1998**, *30*, 159–166.
33. Carraro, G.U.; Cortes, M.; Edmark, J.T.; Ensor, J.R. The peloton bicycling simulator. In Proceedings of the Third Symposium on Virtual Reality Modeling Language, New York, NY, USA, 16–20 February 1998; pp. 63–70.
34. Petrigna, L.; Bianco, A. Is the human body able to travel on Mars? *Sci. Sports* **2022**, *37*, 519–524. [CrossRef]

35. Kostick-Quenet, K.; Mandl, K.D.; Minssen, T.; Cohen, I.G.; Gasser, U.; Kohane, I.; McGuire, A.L. How NFTs could transform health information exchange. *Science* **2022**, *375*, 500–502. [CrossRef] [PubMed]
36. Petrigna, L.; Gentile, A.; Mani, D.; Pajaujiene, S.; Zanotto, T.; Thomas, E.; Paoli, A.; Palma, A.; Bianco, A. Dual-Task Conditions on Static Postural Control in Older Adults: A Systematic Review and Meta-Analysis. *J. Aging Phys. Act.* **2021**, *29*, 162–177. [CrossRef] [PubMed]

MDPI
St. Alban-Anlage 66
4052 Basel
Switzerland
www.mdpi.com

Journal of Functional Morphology and Kinesiology Editorial Office
E-mail: jfmk@mdpi.com
www.mdpi.com/journal/jfmk